Editor
Dona Herweck Rice

Editorial Project Manager
Karen J. Goldfluss, M.S. Ed.

Editor in Chief
Sharon Coan, M.S. Ed.

Illustrator
Wendy Chang

Cover Artist
Larry Bauer

Art Coordinator
Cheri Macoubrie Wilson

Creative Director
Elayne Roberts

Imaging
James Edward Grace

Product Manager
Phil Garcia

Publishers
Rachelle Cracchiolo, M.S. Ed.
Mary Dupuy Smith, M.S. Ed.

Beyond Science Fair

FAMILY SCIENCE

SCIENCE DISCOVERY DAY

INVENTOR'S CONVENTION

Author
Ruth M. Young, M.S. Ed.

Teacher Created Materials, Inc.
6421 Industry Way
Westminster, CA 92683
www.teachercreated.com
ISBN-1-57690-509-8
©1999 Teacher Created Materials, Inc.
Made in U.S.A.

The classroom teacher may reproduce copies of materials in this book for classroom use only. The reproduction of any part for an entire school or school system is strictly prohibited. No part of this publication may be transmitted, stored, or recorded in any form without written permission from the publisher.

Table of Contents

Introduction 3
   A Brief History of the Science Fair 3
   Goals for Conducting an Elementary Grade Science Event 4

Science Discovery Day 5
   Overview 5
   Time Line and Committee Responsibilities .. 7
   Using a Ticket System 12
   Ticket Information Form 15
   Ticket Template 16
   Call for Volunteers 17
   Presenter Form 18
   Confirmation for Presenters 19
   Confirmation for Presiders 20
   Science Discovery Day Map 21
   Science Discovery Day Information 22
   Presenter/Presider Information 23
   Invitation 24
   Sample Parent Letter 25
   Teacher/Presenter Evaluation 26
   K–2 Student Evaluation 27
   Grades 3–6 Student Evaluation 28
   Activity Ideas 29
   Science Discovery Day Challenges 41

Inventors Convention 42
   Overview 42
   Invention Detective 43
   Parent Letter for Invention Detective Assignments 45
   History and Evolution of Inventions 46
   Parent Letter for History of Inventions 48
   Important Inventions in History 49
   Evolution of the Lollipop 51

   History of Inventions 52
   Famous Inventors 53
   Useful Invention 54
   Invent a Musical Instrument 55
   Invent a Tool 56
   Invent a Toy 57
   The Need to Invent 58
   Parent Letter for Physical Challenge Activity 59
   Physical Challenge Cards 60
   Being an Inventor 62
   Ideas for Inventions 63
   A New Invention 64
   Display Instructions 65
   Organizing the Event 66
   List of Inventions for Convention 68
   Inventors Convention Summary 69

Family Science 70
   Overview 70
   Activity Ideas 73
   Center Activities 76
   Family Nature Walks 78
   Sample Nature Walk Announcement 81
   Evening Science 82
   Sample Evening Science Announcement .. 83
   Sample Evening Science Confirmation ... 84
   Sample Waiting List Notice and Scholarship Fund Acknowledgement 85
   Science Performances 86
   Science Performance Ideas 87
   Science Plays 91

Resources 94

Introduction

A Brief History of the Science Fair

The earliest nationwide science contest is the Westinghouse Science Search, begun in 1942. The International Science and Engineering Fair was initiated in 1950. Science fairs were originally conducted for high school students to encourage their interest in pursuing careers in scientific fields. Awards were, and still are, sponsored by many science-related companies and organizations. The best projects selected by a team of judges receive awards that may vary from ribbons to savings bonds. Those winning top awards in local fairs are often given opportunities to enter broader fairs such as those conducted at the state or national levels.

The projects and displays created by high school students are based on scientific research. Sometimes their projects are based on research being conducted by practicing scientists who serve as mentors for students as they prepare for the science fair. Students who team with these mentors usually produce award-winning projects, but they do not always continue in scientific fields. However, the experience of conducting real research is of great value to them.

When elementary teachers began to conduct science fairs, they followed the high school model. These younger students reaped little if any benefit from conducting a research project. These children had neither the scientific skills nor the background knowledge to conduct research projects. Their projects, therefore, frequently were created by older siblings or parents, and thus the elementary student did not learn much science or benefit from the project. Elementary science fairs often put the emphasis on the final write-up of the project; thus, it becomes more a language arts experience than one of science.

Elementary science textbooks, and thereby many elementary science programs, once developed their lessons around the "scientific method." They did not emphasize student discovery of science through exploration but rather used a cookbook method of teaching science.

"Scientific inquiry is more complex than popular conceptions would have it. It is, for instance, a more subtle and demanding process than the naive idea of 'making a great many careful observations and then organizing them.' It is far more flexible than the rigid sequence of steps commonly depicted in textbooks as 'the scientific method.' It is much more than just "doing experiments" (*Benchmarks for Scientific Literacy,* Project 2061 for the American Association for the Advancement of Science, 1993).

The emphasis of teaching science to elementary students has shifted from using the "scientific method" and memorization of a body of scientific knowledge (facts) to developing science process skills and increasing the understanding of scientific concepts through real-life science experiences. This method enables students to discover the wonders of science on a personal level. It also applies and hones their natural skills of curiosity and investigation.

The change in the way K–12 science is taught was firmly established in 1996 when the National Science Education Standards were adopted. These standards emphasize teaching and learning about science that reflects how science itself is done, emphasizing inquiry as a way of achieving knowledge and understanding about the world.

Introduction •

Goals for Conducting an Elementary Grade Science Event

The typical science fair in which elementary students did "research" projects no longer matches the way science is taught. Alternative science events emphasizing interactive science and the enjoyment of doing science are now being used by many elementary schools. The present goals for conducting an elementary science event may include the following:

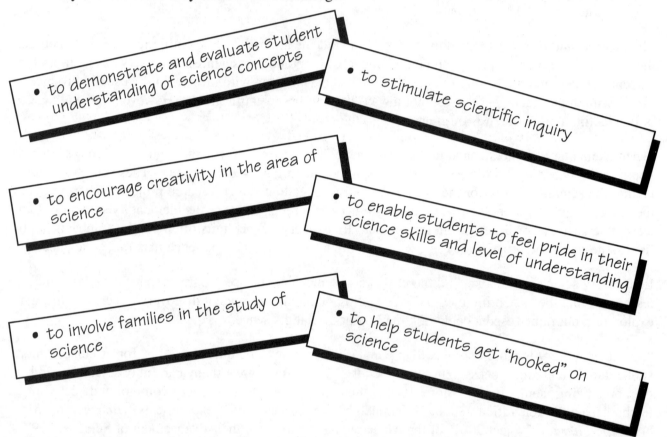

- to demonstrate and evaluate student understanding of science concepts
- to stimulate scientific inquiry
- to encourage creativity in the area of science
- to enable students to feel pride in their science skills and level of understanding
- to involve families in the study of science
- to help students get "hooked" on science

There are a variety of ways to accomplish these and other goals, which will be a far more valuable experience for elementary students than working on a science fair project. Details for conducting a variety of science events which are appropriate for elementary students are provided in this book. They include the following:

Science Discovery Day: a one-day event much like a mini-science conference which provides hands-on activities for all students

Inventors Convention: an opportunity for students to apply science and creative skills by designing a model of a new product

Family Science: a variety of experiences spread across the school year, involving students and adults in science after school hours at the school and other sites

These are but a few ways of bringing science to life at your school. It is hoped that they will lay the groundwork for science events in which the entire school population can become involved for the benefit of students, staff, and families.

Science Discovery Day

Overview

Science Discovery Day is an event which offers a wide variety of science activity sessions for students to attend. It is structured much like a mini-science conference. One or two teachers and/or school administrators should serve as coordinator(s) who will assume the responsibility of organizing the event, including overseeing committees that handle many of the day's details. Presenters of the activities may be teachers, parents, upper-grade students, and guests from local museums, zoos, colleges, and hospitals. This event is designed for grades K–6, but students from middle and high school may serve as presenters.

Science Discovery Day should involve everyone at the school so that all rooms and outdoor areas are available for sessions. Each presenter will repeat his or her activity three or four times, depending upon the number of sessions scheduled. Schedule the sessions for 40–45 minutes with a 10 minute passing period. This will give the presenters time to involve students in an exciting science activity and then clean up and prepare for the next group. Scheduling recess for students and a break for presenters between the second and final session(s) provides a much needed rest period. This is also a good time to serve light refreshments for the hard-working presenters and assistants.

The activities should be assigned to different grade levels according to the nature of the activities and the preference of the presenters. There should be enough activities to spread the students into groups of no more than fifteen, making it easier for the presenter. The entire school site may be required to accommodate this event, including every classroom and shared building as well as the outside areas. Schedule Science Discovery Day during the season of the year which is most likely to provide good weather. If the date is near the end of the year, it may be the grand finale for all the science programs the students have experienced since the beginning of school.

Student Presenters

Fifth and sixth grade students may be included as presenters. It is best to develop teams of six students for each activity so they can rotate in pairs of presenters for each of the three sessions. This will make it possible for each team member to attend two other activities during the day. The teams may develop their own activities, or one may be assigned to them. The student presenters will need training and an opportunity to practice. An adult presider should be assigned to each student-led activity throughout all three sessions. Be sure to explain to the presiders that they will assist the presenters but are not to take over the instruction. Students may also preside, if desired, perhaps working with their teacher.

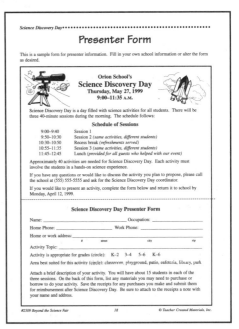

Consider asking junior and senior high school science teachers to allow some of their students to develop an activity for presentation. They should complete a Presenter Form (page 18) so their activity description will be provided, as well as any equipment or special site for their sessions. Request that the science teacher provide the opportunity for their student(s) to practice their presentation before their peers prior to conducting it at your school.

Science Discovery Day

Overview (cont.)

Screening Activities

It is a good idea to screen the activities proposed by off-site presenters to be sure they will actively involve the students rather than lecture, merely demonstrate, or only a display. If someone proposes an activity which does not involve the students, the coordinators may need to work with the presenter to show how it can be done. This usually takes a bit of diplomacy, but the presenter will benefit by discovering that teaching science requires skills in creating lessons that have students learn through participation. For example,

- viewing a mineral collection will not involve the students as much as showing the minerals and then leading the students through an activity that helps them learn to identify minerals.

- showing a video on protecting wildlife can be extended into an activity by showing actual wild animals which have been rescued and letting students perform a rescue simulation. Contact the local humane society to request a speaker and animal expert for this topic.

- juggling can begin as a demonstration but should quickly involve the students in juggling.

Organizing the Event

Science Discovery Day is a schoolwide effort and needs coordination and long-range planning to make it a success. Begin this planning at least three months before the date of the event. This date should be determined by the teachers since they will play the most important roles.

One or two people should serve as coordinator(s) who will be responsible for establishing and overseeing committees. A teacher and/or administrator from the school should serve as coordinator(s). The coordinators should not be responsible for presenting any sessions since they will be needed to make sure everything is going smoothly throughout the day of the event.

The information on the following pages will assist you in organizing this special day. A suggested time line begins on the next page. The time line and forms should be modified to fit the needs of your school. A description for using a ticket system to rotate students can be found on pages 12–16. Sample forms have been provided on pages 17–28. Examples of activities are briefly described on pages 29–40.

Time Line and Committee Responsibilities

❑ **Three Months Before the Event**

Establish Committees

Committees should be formed to spread the work load among many people. The number of members on each committee will vary, as well as the amount of time needed to complete the tasks. Committees should include administrator(s), teachers, parents, and, possibly, students from the school. The committees and some of their responsibilities are described below.

Program Committee

- Determine the number of activities needed to have a maximum of 15 students each.
- Solicit presenters to do activities and presiders to help the presenters.
- Screen presenter applications to be sure activities are appropriate and involve students.
- Check to be sure there are activities for all grades to accommodate everyone.
- Assign each presenter to an appropriate classroom or outdoor area.
- Assign presiders to each activity to act as assistants to the presenter.
- Create a list of the activities and their locations.
- Make a school map showing where each activity will be located.
- Make name tags for presenters, presiders, and other personnel who will be involved.
- Make signs with the title, grade level(s), and presenter's name to be posted at the sites.

Public Relations Committee

- Disseminate information about the event to families.
- Issue invitations to board members, district office administrators, and principals.
- Notify the media about the event and send them invitations to attend.
- Arrange to have volunteers videotape and photograph the event. Make these available after the event. These visual records may also be used for a presentation to the PTA, school board, or other schools which may want to conduct their own Science Discovery Day.

Materials and Equipment Committee

- Create a list of equipment and materials needed for each activity.
- Gather equipment and purchase the materials designated by each presenter.
- Make a school map showing where equipment and materials are needed.
- Distribute materials and equipment to the activity sites on the day prior to the event (or two hours prior if the event will take place outside).
- Assign a person to locate additional materials and equipment as needed throughout the day.
- Establish a system for the return of equipment and materials at the end of the day.
- Check the returned items with the list to be sure all have been collected and then redistribute them to their original locations.

Science Discovery Day

Time Line and Committee Responsibilities (cont.)

Facilities and School Grounds Committee

- Arrange for student supervision during periods between sessions and at the recess break.
- Arrange for escorts to help special education and/or younger students move between activity locations. Older students may be included as escorts.
- If the event will alter the regular lunch schedule, notify cafeteria personnel at least one week in advance. Arrange for students to eat together, and encourage them to bring a lunch to decrease demands on the cafeteria. Consider eating outdoors to relieve cafeteria congestion.
- Check all activity sites the week before the event to be sure they are ready for use.
- Make any necessary preparations to assist visiting presenters with such things as parking and transporting materials to their sites.
- Organize a system for rotating students through the activities, perhaps by issuing tickets for each activity and session.
- Check all activity locations after the event to be sure they are in good condition.

Hospitality Committee

- Arrange for greeters and tour guides to assist on the day of the event.
- Arrange for a table at the school entrance where greeters will sign in and give name tags to presenters and presiders. Include visitor information and maps.
- Arrange for refreshments during the recess break.
- Consider providing a potluck lunch at the end of the day for all volunteers, with food provided by the PTA or teachers.
- After the event, send notes of appreciation to all helpers.

Evaluation Committee

- Design evaluation forms, one each for teachers, students, and presenters. Distribute the forms to teachers and presenters before the event. Teachers can give them to their students at the end of the day to complete and return before dismissal.
- Arrange for a central collection box where completed forms can be deposited.
- Make a summary report of the evaluations for the coordinators.

Design and Distribute Forms

Forms will be needed to attract volunteers and presenters as well as to evaluate the event. The volunteer forms can be distributed to all families at the school as well as specific locations such as local museums, hospitals, industries, colleges, and others you feel can offer quality activities. The presenters form can also be sent to those who indicate an interest in presenting on their volunteer forms. Be sure to give each teacher a presenter form if he or she plans to lead an activity. Forms are included on pages 15 through 28. These can be modified as needed for your school.

Time Line and Committee Responsibilities (cont.)

❑ Two Months Before the Event

- All presenter and volunteer forms should be given to the Program Committee.
- Assign activities to appropriate grade levels and sites.
- Assign a presider to each presenter.
- Create a map of the school grounds, that shows the location of each activity (sample on page 21). Make several enlarged copies of the map to be posted on the grounds during the event.
- Send confirmations to volunteers (sample on page 17). Include the map on page 21.
- Make a master list of all activities, presenters, assistants, grade levels, and locations (samples on pages 22 and 23).

❑ One Month Before the Event

- Purchase or locate necessary materials.
- Arrange for greeters for the event.
- Arrange for tour guides to take visitors to the sessions (adults or older students).
- Design a student rotation system for the event. A ticket system is provided on pages 12–14.
- Arrange for a photographer to record each activity briefly for a history of the event.
- Decide how to advertise the event to the parents and public. This can be done by distributing flyers and posting a banner outside the school just prior to the event.
- Make arrangements for refreshments which may be served to the presenters and volunteers during the break. If food is to be served afterwards, also make arrangements for this.

❑ Three Weeks Before the Event

- Make arrangements for supervision of students during the recess break. Teachers who are presenting sessions will need time to prepare for the final session and thus be unavailable for this duty. If lunch is being served to the staff and guests, make arrangements with the cafeteria and arrange for student eating areas as well as their supervision during lunch.
- Prepare a map of the school showing activity locations, with a list of the activities, including grade levels, presenter names, and other information printed on the back of the map.
- Notify the media about Science Discovery Day and invite press coverage. (Check school policy before arranging this event.)
- Send invitations to special guests such as the superintendent and principals of other schools.
- Arrange for refreshments to be served at the break.
- Schedule greeters and tour guides to be available at the visitor table to assist visitors throughout the day. Give a copy of the schedule to each greeter, tour guide, and the school secretary.
- Provide office staff with copies of the map, the list of activities and presenters, and other information which may be needed during the event.

Time Line and Committee Responsibilities *(cont.)*

❑ One Week Before the Event

- Deliver the materials needed by each presenter to the classrooms. Save materials being used at outdoor activities to deliver on the day of the event.
- Arrange for the location of refreshments that will be available for presenters and presiders at the break. Be sure to announce the availability of refreshments in their confirmations.
- Contact media representatives to remind them of the event and to reissue invitations.
- Contact off-site presenters to remind them of the date and schedule. Ask if they need any assistance such as special parking for unloading. Check that they have all supplies.
- Make name tags for presenters and presiders. Distribute these to on-site people. Name tags for guest presenters and presiders should be displayed at the check-in table.
- If a ticket system is being used to rotate students, distribute the tickets to teachers.
- Teachers should arrange escorts for students who may need assistance finding their rooms. Kindergarten children can be placed in groups of fifteen so they can move together with one escort.
- Teachers make name tags for each of their students.
- Teachers assign tickets to the students and staple them into sets of three.

❑ Two Hours Before the Event

- Teachers distribute name tags and tell students to wear them so they can be seen by presenters and others who need to know their names.
- Teachers distribute ticket sets to the students.
- Near the entrance of the school place a table which will have a Science Discovery Day map and a printed list of the activities printed for visitors. Include a guest book if you like, as well as name tags for guests to make for themselves. Greeters and tour guides should be stationed in this area so they can assist visitors as needed.
- Deliver the materials to outdoor sites.
- Post the enlarged copies of the Science Discovery Day map around the grounds.
- Post the sign for each activity on the door or near its location.

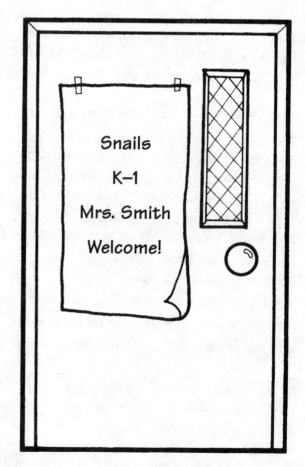

Time Line and Committee Responsibilities (cont.)

❑ One Hour Before the Event

- Coordinators should make a final check of each activity site to be sure all presenters are in place and have the materials they will need.

❑ During the Event

- Coordinators circulate to be sure all is running smoothly. They should visit each activity to witness the excitement students are feeling as they discover that science is fun!

❑ Before Dismissal

- Students will have attended a variety of activities. They should have time to share within their own classrooms what they did in their sessions before going home.
- Have the students, volunteers, and presenters complete an evaluation form to find out how they felt about the day and to gather suggestions for changes which may be needed when this event is conducted in the future.
- Revel in the great comments of students, volunteers, and presenters. You will know the event has been a huge success when many students say, "Let's do it again tomorrow!"

❑ After Science Discovery Day

- When the pictures are developed, display them for everyone at the school to see. Let teachers show the videotape to their classes so they can see samples of the many activities which went on throughout the day.
- Send thank-you notes to off-site volunteers. Science postcards may be used for this purpose.
- Coordinators should conduct a "debriefing" meeting with all the committees to review the evaluations, pictures, and general feelings regarding the success of the Science Discovery Day.
- Develop plans for repeating this exciting event in the future. Write guidelines based on what you learned this year to help assist the coordinators in the future.

Science Discovery Day •

Using a Ticket System

One method of rotating students through the activities is to use a ticket system. Assuming that there will be three sessions, each student will receive three tickets, one for each session. A master for the tickets should be made and then printed on red, green, and yellow paper (the traffic light colors to move students through the three sessions). First session tickets would be printed on green, second session on yellow, and third session on red. A template for making the tickets is provided on page 16. Fill in the missing information for each activity, including its number. Once all ticket information is included on the template, print 15 copies on each of the three colors to be used to indicate each session. An example of the ticket is shown below. The activity number is shown on the left side of the ticket. This ticket is for activity number 3.

3	Activity: Snails Grade(s): K–1 Location: Kindergarten Presenter: Mrs. Smith

Sample Science Discovery Day Ticket

To determine the number of participants from the designated grade levels for each activity, complete the Ticket Information Form on page 15. An example is shown below. Notice that activities are limited to 15 participants.

Ticket Information Form

#	Grades	Activity Title	K	1	2	3	4	5	6	Total
1	K–1	What Do You Hear?	10	5						15
2	K–2	Animal Crackers	5	5	4					14
3	K–2	Guessing Jar	10	2	3					15
8	2–3	Building Arches			5	7				12
9	2–3	Colors Galore			5	6				11
15	3–4	Parachutes				6	9			15
16	3–4	Potato Eyes				7	8			15
20	4–6	Building Bridges					5	5	5	15
24	5–6	Space to Earth						9	5	14
Total Students in School			56	74	78	93	109	93	93	596

#2509 Beyond the Science Fair   © Teacher Created Materials, Inc.

Using a Ticket System (cont.)

Instructions for Completing the Ticket Information Form

(The form can be found on page 15.)

1. Beginning with 1, place numbers in the # column. These are the ticket numbers. Make additional copies of this form to include all activities.

2. List the activities in order of grade level, beginning with kindergarten. Alphabetize the titles within each grade level block as shown on the form example (page 12).

3. Two weeks prior to the event, complete the "Total Students in School" and "Total" sections of the Ticket Information Form.

4. Determine the number of students who fall into each grade level and divide them among the activities for their grades. To make the task simpler, keep the number of participants per grade level the same for each of the three sessions.

5. The count of students per classroom should be checked prior to ticket distribution to adjust for students who have left the school or recently enrolled. When possible, assign new students to the activities with fewer than 15 participants.

Distributing the Tickets

Follow the steps below to prepare the tickets for distribution to each teacher. Those doing this task should work at a long table.

1. Count the number of tickets to be allocated for each activity by grade level as shown on the Ticket Information Form. For example, using the sample form (page 12), the K–1 activity "What Do You Hear?" has a total of 15 tickets, 10 for kindergarten, and five for first grade. "Animal Crackers" will have five for kindergarten, five for first grade, and four for second grade. This number of tickets remains the same for all three sessions.

2. Count the number of green tickets for each activity designated for one grade level (e.g., kindergarten). Bundle the set of tickets by number and secure them with a rubber band. Each ticket bundle will contain all the tickets for all activities designed for the grade level. The total in each bundle should equal the number of students in that grade level. There will be extra tickets for activities which do not have a total of 15 participants. These should be set aside and used for new students who may arrive once tickets have been distributed to the classrooms.

3. Repeat the bundling process with the yellow and red tickets for kindergarten.

4. Use this same procedure for all the grades until all tickets have been bundled.

Science Discovery Day •

Using a Ticket System (cont.)

Distributing the Tickets (cont.)

5. Label a file card for each class with the teacher's name and total number of students in that classroom. See the example below.

> Classroom #:   19
> Teacher: Nelson
> Grade 3:   22

When a classroom has students from more than one grade level, make separate file cards for each grade. Note the example below.

> Classroom #:   3   Classroom #:   3
> Teacher: Jones   Teacher: Jones
> Grade 5:   15   Grade 6:   10

6. Lay the file cards for one grade level on a table, beginning with kindergarten.
7. Begin with green tickets for kindergarten and then deal them like playing cards in piles beneath the classroom file cards. Allocate sufficient tickets so each student in the class will get one. Deal the green tickets until all have been distributed. Band these together as a set for each kindergarten classroom and lay them below the card. Dealing the tickets in this manner will ensure that students in each classroom attend a variety of activities.
8. Continue dealing the yellow and then red tickets for kindergarten until all have been distributed and banded.
9. Count the tickets for each classroom to be sure there is the correct number of tickets in each color for the number of students in the class. Place each classroom card and sets of tickets into an envelope marked with the teacher's name.
10. When kindergarten is complete, repeat the process for each of the next grades. Teachers with more than one grade level should receive a separate envelope for each.
11. Deliver the ticket sets to the teachers one week before the event.

Teacher Allocation of Tickets

After teachers receive their tickets, they are responsible for distributing them to their students. Each student will need a different activity ticket for each of three sessions. The teacher can determine how the students receive their tickets. They may wish to let students choose one ticket and then assign the rest of them. This provides the opportunity for each child to have at least one choice. Before the tickets are distributed to the students, they should be stapled together according to the color code system for rotation (green on top, yellow in the middle, and red at the bottom). Check to be sure all three tickets are for different activities before stapling them.

Ticket sets should be distributed to the students on the morning of the event to be sure they are not misplaced. Younger students may need to have their ticket sets pinned or taped to their clothing to be sure they do not become lost during the day. If kindergarten students are grouped together, their tickets may be given to their escorts. Tickets should be checked and collected by the presider at each activity.

Science Discovery Day

Ticket Information Form

Ticket Information Form									
#	Grades	Activity Title	K	1	2	3	4	5	6	Total
	Total Students in School									

Science Discovery Day •

Ticket Template

	Activity: Grade(s): Location: Presenter:		Activity: Grade(s): Location: Presenter:
	Activity: Grade(s): Location: Presenter:		Activity: Grade(s): Location: Presenter:
	Activity: Grade(s): Location: Presenter:		Activity: Grade(s): Location: Presenter:
	Activity: Grade(s): Location: Presenter:		Activity: Grade(s): Location: Presenter:
	Activity: Grade(s): Location: Presenter:		Activity: Grade(s): Location: Presenter:
	Activity: Grade(s): Location: Presenter:		Activity: Grade(s): Location: Presenter:
	Activity: Grade(s): Location: Presenter:		Activity: Grade(s): Location: Presenter:

Call for Volunteers

This is a sample form for soliciting volunteers. Fill in your own school information or alter the form as desired.

Orion School
Announces a Special Event
Science Discovery Day
Thursday, May 27, 1999
9:00–11:35 A.M.

Science Discovery Day is a day filled with science activities for all students at the school. There will be three 40-minute sessions during the morning. Approximately 40 activities will be presented during each of the sessions. The schedule follows:

Schedule of Sessions

9:00–9:40	Session 1
9:50–10:30	Session 2 *(same activities, different students)*
10:30–10:50	Recess break *(refreshments served)*
10:55–11:35	Session 3 *(same activities, different students)*
11:45–12:45	Lunch *(provided for all guests who helped with our event)*

We need your help to make our Science Discovery Day a success. If you are interested in participating as a volunteer presenter to do an activity, to serve as an assistant to a presenter, or to help on the grounds during the event, please complete the form below. If you have any questions, please call the school at (555) 555-5555 and ask for the Science Discovery Day coordinator.

Please tear off the form below and *return it to school by Friday, March 26, 1999.*

- -

Science Discovery Day Volunteer Form

Please print the following information:

Name: _____ Phone: _____

Child's Name: _____ Child's Teacher: _____

_____ I would like to be a presider and assist a presenter with his or her activity during all three sessions.

_____ I would like to assist as needed on the school site from 8:45–11:45.

_____ I would like to present a science activity. (**Note:** A presenter form will be sent to you.)

Science Discovery Day

Presenter Form

This is a sample form for presenter information. Fill in your own school information or alter the form as desired.

Orion School's
Science Discovery Day
Thursday, May 27, 1999
9:00–11:35 A.M.

Science Discovery Day is a day filled with science activities for all students. There will be three 40-minute sessions during the morning. The schedule follows:

Schedule of Sessions

9:00–9:40	Session 1
9:50–10:30	Session 2 *(same activities, different students)*
10:30–10:50	Recess break *(refreshments served)*
10:55–11:35	Session 3 *(same activities, different students)*
11:45–12:45	Lunch *(provided for all guests who helped with our event)*

Approximately 40 activities are needed for Science Discovery Day. Each activity must involve the students in a hands-on science experience.

If you have any questions or would like to discuss the activity you plan to propose, please call the school at (555) 555-5555 and ask for the Science Discovery Day coordinator.

If you would like to present an activity, complete the form below and return it to school by Monday, April 12, 1999.

- -

Science Discovery Day Presenter Form

Name: _____ Occupation: _____

Home Phone: _____ Work Phone: _____

Home or work address: _____
                         #   street   city   zip

Activity Topic: _____

Activity is appropriate for grades (circle):   K–2   3–4   5–6   K–6

Area best suited for this activity (circle): classroom, playground, patio, cafeteria, library, park

Attach a brief description of your activity. You will have about 15 students in each of the three sessions. On the back of this form, list any materials you may need to purchase or borrow to do your activity. Save the receipts for any purchases you make and submit them for reimbursement after Science Discovery Day. Be sure to attach to the receipts a note with your name and address.

Science Discovery Day

Confirmation for Presenters

This is a sample confirmation form for presenters. Fill in your own school information or alter the form as desired.

Orion School's
Science Discovery Day
Thursday, May 27, 1999
9:00–11:35 A.M.

Presenter: _____ **Presider:** _____

Topic: _____ **Presentation Site:** _____

We are delighted that you will be a presenter at this year's Science Discovery Day. To help you locate your activity site, a map of the school grounds is attached, showing the locations of all the activities being presented. The following information regards your presentations during the event.

Schedule of Sessions

9:00–9:40	Session 1
9:50–10:30	Session 2 *(same activities, different students)*
10:30–10:50	Recess break *(refreshments served in Teacher's Lounge)*
10:55–11:35	Session 3 *(same activities, different students)*
11:45–12:45	Lunch *(provided for all guests who helped with our event)*

Please plan to arrive by 8:30 A.M. and check in at the hospitality table in front of the school. There you will receive your name tag and any assistance you may require to transport your materials to your presentation site.

You have requested to borrow the materials listed below. They will be at your site when you arrive. If you have any changes to make, please notify the Science Discovery Day coordinator at the school as soon as possible by calling (555) 555-5555.

Materials Requested:

_____ _____ _____

_____ _____ _____

Remember, if you purchase any consumable materials for use in your sessions, please submit the receipt for reimbursement. The receipts should be attached to a note which gives your name and address. Submit these to the coordinator on the day of the event or no later than one week after it.

Important: If for any reason you are unable to present your session, please notify the school as soon as possible. The students are assigned by tickets to each of the events; therefore, if one is cancelled, it will be necessary to reassign the attendees.

Science Discovery Day

Confirmation for Presiders

This is a sample confirmation form for presiders. Fill in your own school information or alter the form as desired.

Orion School's
Science Discovery Day
Thursday, May 27, 1999
9:00–11:35 A.M.

Presenter: _____   **Presider:** _____

Topic: _____   **Presentation Site:** _____

We are delighted that you will be a presider at this year's Science Discovery Day. To help you locate your activity site, a map of the school grounds is attached, showing the locations of all the activities being presented. The schedule for the sessions is shown below.

Schedule of Sessions

9:00–9:40	Session 1
9:50–10:30	Session 2 *(same activities, different students)*
10:30–10:50	Recess break *(refreshments served in Teacher's Lounge)*
10:55–11:35	Session 3 *(same activities, different students)*
11:45–12:45	Lunch *(provided for all guests who helped with our event)*

The following information is in regard to your duties as a presider during this event.

- Please plan to arrive by 8:30 A.M. and check in at the hospitality table in the front of the school to pick up your nametag.

- Report to the site of the activity for which you will preside, as shown on the attached map. Introduce yourself to the presenter and ask what you can do to help him or her.

- As a presider, you will be assisting the presenter throughout the three activity sessions. This may include helping to distribute materials and cleaning up between sessions. You should also assist in cleaning up at the area after the final session.

- As students come to the activity, collect their tickets and be sure the activity listed on the ticket is the one taking place at this location. If it is not, use the map to help the student go to the right area. Also, check the color of the ticket to be sure it is for the right session. The tickets are color coded: green tickets are for the first session, yellow for the second, and red for the last. The tickets may be discarded at the end of the event.

Optional: for each presider's activity, attach a sample ticket to the confirmation form.

Important: If for any reason you are unable to attend, please notify the Science Discovery Day coordinator(s) at the school as soon as possible by calling (555) 555-5555. We will assign another volunteer in your place since we want all presenters to have presiders assist them.

Science Discovery Day

Science Discovery Day Map

This is a sample map. Use it as a model to create your own map.

Orion School

Kites and Juggling *(lower playground)*   **Discovery Walk** *(park)*

| Room 22 Lemons to Lemonade | Room 21 Optricks | Room 20 Signal Flags | Room 19 Owl Pellets | Room 18 Pets | Restrooms |

Bubbles *(sandbox area)*

| Room 17 Gliders | Room 16 Airplanes | Room 15 Go for Launch | Room 14 Electric Circuits | Room 13 Colored Solutions |

Geology *(patio)*

| Restrooms | Room 12 Finger Paints | Room 11 Bridges | Room 10 Chemical Magic | Room 9 Structures |

| Room 8 Crystal Gardens | Room 7 Magnets | Room 6 Show & Tell | Room 5 Sorting | Room 4 Flubber |

Office

Lounge

K Classroom

Good Vibrations

Parachutes *(upper playground)*

| Room 3 Fish Dissection | Room 2 Mealworms | Room 1 Dry Ice Fossils |

Schedule

| Session 1 | 9:00–9:40 | Recess break | 10:30–10:50 | Lunch | 11:45–12:45 |
| Session 2 | 9:50–10:30 | Session 3 | 10:55–11:35 | | |

© Teacher Created Materials, Inc.

Science Discovery Day •

Science Discovery Day Information

#	Grade	Activity Title	Site	Presenter	Presider	Equipment/Materials

Science Discovery Day

Presenter/Presider Information

#	Activity Title	Presenter	Phone	Presider	Phone

Science Discovery Day •

Invitation

This is a sample invitation to Science Discovery Day. Fill in your school's information or use this as a model to create your own invitation.

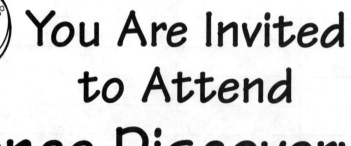

You Are Invited to Attend
Science Discovery Day

Thursday, May 27, 1999

9:00–11:35 A.M.

Science Discovery Day is a special event that involves all students in a variety of science activities.

Schedule of Sessions

9:00–9:45	Session 1
9:50–10:30	Session 2
10:30–10:50	Recess Break
10:55–11:35	Session 3

Come Join the Fun!

Please stop by the hospitality table at the school entrance to pick up a map of the school, which will show all activity locations.

If you have any questions, please call the Science Discovery Day coordinator(s) at (555) 555-5555.

#2509 Beyond the Science Fair — © Teacher Created Materials, Inc.

Sample Parent Letter

This is a sample letter to parents regarding Science Discovery Day. Fill in your school's information or use this as a model to create your own letter.

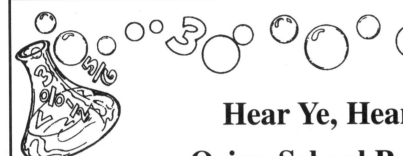

Hear Ye, Hear Ye
Orion School Presents
Science Discovery Day

Thursday, May 27, 1999

9:00–11:35 A.M.

Science Discovery Day is a special event that involves all students in a variety of science activities. As always, parents are welcome to observe.

Schedule of Sessions

9:00 - 9:40	Session 1
9:50 - 10:30	Session 2
10:30 - 10:50	Recess Break
10:55 - 11:35	Session 3

Students will eat lunch at the same time on this day. Although lunch will be served as usual, it will be helpful if students bring a sack lunch to help avoid crowding in the lunch line.

If you have any questions, please call the Science Discovery Day coordinator(s) at the school at (555) 555-5555.

Science Discovery Day •

Teacher/Presenter Evaluation

The evaluation committee needs your help in evaluating Science Discovery Day. Please complete the form below, giving us your assessment. When completed, leave the form in the evaluation box located in _____. The form must be in the box by _____ in order to be included in the committee's summary.

Name (optional): _____

Please list the best features of Science Discovery Day:

_____
_____
_____
_____
_____
_____
_____

Please list suggestions for improving Science Discovery Day:

_____
_____
_____
_____
_____
_____
_____

Would you like to have Science Discovery Day offered next year?   Yes   No

If no, please explain why:

_____
_____
_____
_____
_____
_____
_____

Thank you for taking the time to complete this form. All evaluations will be summarized, and a report will be made for the Science Discovery Day coordinator(s) and school staff.

•• *Science Discovery Day*

K–2 Student Evaluation

The evaluation committee needs your help in evaluating Science Discovery Day. Please list the activities your students attended today and then ask them the questions below. Summarize their answers and write them here. Feel free to use both sides of the form. When completed, leave the form in the evaluation box located in _____. The form must be in the box by _____ in order to be included in the committee's summary.

Teacher's Name: _____

Grade: _____

List the activities your students attended today:

_____
_____
_____
_____
_____
_____
_____
_____
_____
_____
_____
_____

What science activities did the students like best?

_____
_____

What percentage of the students would like to see Science Discovery Day offered again next year?

_____

If any students would not like, what are their reasons?

_____
_____

What can we do to make Science Discovery Day better?

_____
_____
_____

Thank you for taking the time to complete this form. All evaluations will be summarized, and a report will be made for the Science Discovery Day coordinator(s) and school staff.

© Teacher Created Materials, Inc.

Science Discovery Day••

Grades 3–6 Student Evaluation

The evaluation committee needs your help in evaluating Science Discovery Day. Please list the activities you attended today and then answer the questions.

Teacher's Name: _____

Grade: _____

List the activities you attended today:

_____
_____
_____
_____
_____
_____

Which activities did you like best and why?

_____
_____
_____
_____
_____
_____

Would you like to see Science Discovery Day offered again next year?
_____

If you replied no, please explain why.
_____
_____
_____

What can we do to make Science Discovery Day better?
_____
_____
_____
_____

Thank you for taking the time to complete this evaluation form. Please return this form to your teacher.

Science Discovery Day

Activity Ideas

Activities for Science Discovery Day can be found in many places. Begin with the science concepts included within each grade's curriculum. Your own books and resources are sure to prove invaluable. Additionally, many of the resources listed on pages 94–96 of this book include activity ideas. Finally, the activities below and through page 40 may also be used.

❑ Grades K–2

Animal Cracker Count

Goal: Graph the various types of animals found in a box of animal crackers.

Materials: box of animal crackers, graph paper, pencil or crayon

Procedure: Each student receives a box of animal crackers. They sort these according to the type of animal. A bar graph is made, showing the number of each type of animal in that box. Add the total number of animal crackers in the box and record it on the graph.

Closure: Compare the graphs to see if the type and quantity of animals represented in the boxes show a pattern.

Boats

Goal: Create a simple boat and check to see how much cargo it can hold.

Materials: sandwich-sized aluminum foil, wading pool, water, cargo (e.g., marbles)

Procedure: Students make boats from aluminum foil and float them in the wading pool. They add cargo one piece at a time to see how much their boats can hold before sinking. Each student gets three trials, making a new boats from another piece of foil each time. This will enable them to redesign their boats to improve them.

Closure: Compare which boats could hold the most cargo and discuss how their shapes made them work better than the boats which held less.

Colored Solutions

Goal: Observe what colors result when food coloring is added to water.

Materials: red, green, and blue food coloring in dropper bottles, clear plastic tumblers, water, toothpicks

Procedure: Divide the students into groups of two or three and give each group a set of coloring bottles and a tumbler of water. Tell the students to put two drops of red food coloring into their water. Observe what happens. Let them discuss how the color fell into the water and spread out in a swirling motion. Have them add two drops of blue to see if it does the same thing. (It does.) Now have them use the toothpick to stir the water. Hold up the cup to look through it and to see what color results. Dump out the water and refill the cup. This time have them add blue and green food coloring, mixing it to see what color results. Finally, mix red and green.

Closure: Ask the students what color they think will result when they mix all three colors. Add two drops of each color to clean water, stir, and see the results.

© Teacher Created Materials, Inc.

Science Discovery Day

Activity Ideas (cont.)

Dancing Spaghetti

Goal: Investigate simple chemical reactions.

Materials: clear carbonated beverage, thin spaghetti, raisins, clear plastic tumblers

Procedure: Pour the carbonated beverage into the plastic tumblers. Let some of the students drop in several pieces of spaghetti while others use about five raisins. Let them watch to see what happens. Encourage them to look for what might be causing the spaghetti and raisins to rise, sink, and rise again.

Have the students discuss what they see in the liquid. Be sure they notice the small bubbles in it.

Closure: Discuss the students' observations. Have them watch carefully as the spaghetti and raisins rise. Students should notice that the foods have bubbles supporting them. As the bubbles rise to the top, have the students notice that the bubbles burst; thus, gravity takes over and the foods sink.

Discovery Walk

(This activity may also be conducted by older students.)

Goal: Walk around the school area, observing things of nature.

Materials: spray bottle of water, small magnifier

Procedure: The presenter should search the grounds several days before the event for locations of spider webs, animal holes, insects, nests, and other interesting examples of nature. On the day of the event, students are led on a route that will take them close to these examples. Students should have magnifiers so they can closely examine some of the things they see. Spray water on any spider webs to make them more visible to the students.

Closure: Have students discuss what they saw and why these things are important to us and to the world.

Kites

Goal: Use a kite to learn about air.

Materials: light plastic kites, 100 feet (30 m) of heavy string, 6" (15 cm) tube of plastic

Procedure: Assemble a kite for each pair of students. Attach the string to the kite and the center of the tube, wind the string around the tube. Take the students to an open field with no power lines or obstructing trees. Have the students line up in pairs across one end of the field. They should face each other with one student holding the kite and the other the tube of string. The student holding the kite holds it high and releases it when his or her partner begins to run down the field. This partner lets the string unwind from the tube as he or she runs until reaching the end of the field. This is repeated with the partners changing places.

Closure: Discuss how the kites flew, what kept them up, and what techniques worked best for keeping them aloft.

Activity Ideas (cont.)

Oobleck

Goal: Students will investigate the properties of solid and liquids.

Materials: *Bartholomew and the Oobleck by* Dr. Seuss, cups of water with a dropper in each, one ounce (25 g) cups one-quarter full of cornstarch, small spoons, green food coloring

Procedure: Read the story about oobleck and then tell students they are going to make it. Divide the students into small groups and give each child a cup of cornstarch. Have the children examine the cornstarch, feeling and smelling it. Let them decide whether it is a solid, liquid, or gas. Give each group a cup of water and dropper. Have them examine the water and tell if it is a solid, liquid, or gas. Tell them to add a few drops of water to the cornstarch until it makes a thick paste so that when it is stirred it is solid, and when left to rest it becomes semi-liquid. Have them add two drops of green food coloring to the mixture and stir until it is mixed. Tell them that they have just made "oobleck."

Closure: Tell the students to each put a spoonful of the mixture into their palms and knead it with their fingers to see how it turns from solid to liquid and back to solid.

Snails

Goal: Learn about the physical characteristics of snails.

Materials: snails, pieces of clear glass taped around the edges, sand in a cake pan, file cards, pencils, hand lenses, lettuce

Procedure: Distribute a snail to each student, along with a hand lens, pencil, and file card. Let them each draw what their snails look like. Tell them to make the picture as big as the card, using the hand lens to look for details. Have the students place their snails on a glass and look underneath to see what it looks like. Tell them to draw what it looks like on the other side of the file card. Let the students experiment with the snail by letting it crawl on a smooth surface and then in the pan of sand. Have them feed the snails lettuce to see them eat. They should observe this with their hand lenses.

Closure: Ask the students to share their pictures and tell what they found out about snails.

The Sorting Game

Goal: Learn how to classify a variety of items.

Materials: sets of various items such as buttons, colored fabrics, geometric shapes, seeds

Procedure: Tell the students you are going to show them how to sort things, using the people in the room. Call up students by name, asking boys to stand in one area and girls in another but do not let the students know this is the method of sorting. After doing this for a while, ask the students to explain how you sorted them. Sort the children by another way, this time perhaps by color of hair. Again ask how they were sorted. Divide the students into groups and provide each with a set of items. Explain that they are to sort the items, using only one characteristic, just as you did when sorting them. Once a group has sorted the items, let them go to another group's area and see if they can determine how the items were sorted. Let the students sort the new items at their area, using other charactcristics.

Closure: Let the students select a way to sort the children in the room and then have them move into groups according to the characteristics they have chosen.

Science Discovery Day

Activity Ideas (cont.)

What Do You Hear?

Goal: Guess what is inside a container from its sound and weight.

Materials: 15 empty, opaque film canisters; small items such as marbles, paper clips, sand

Procedure: Place small object(s) into each canister, such as two marbles. Number each container and make a list of its contents. Make an answer sheet which lists at random the items in the 15 containers. Let students shake and listen to each container. They can write the number of the container next to the matching item on the answer sheet.

Closure: Open the containers and let students check their answers. Discuss how they decided what was inside.

❑ Grades 3–4

Building Bridges

Goal: Students will construct bridges from drinking straws and test their strength.

Materials: drinking straws, clear tape, scissors, metric masses, pictures of a variety of bridges

Procedure: Show the students the pictures of bridges and discuss what they look like. Explain that they are now going to be bridge engineers and will work in small groups to design and construct a bridge from straws. Tell them that each group will have the same number of straws and the same amount of time to build the bridge. Let them know that each bridge will be tested for strength after it is built. Divide the students into groups and distribute an equal number of straws, a roll of clear tape, scissors, and metric masses to pre-test their bridges. Write the time on the board and let them begin to build. Give a 10-minute and then a five-minute warning before they are to stop.

Closure: Let each group test its bridge for strength to see which design is the sturdiest. Have them compare their bridges with the pictures shown at the beginning.

Constructions

Goal: Construct a variety of buildings and equipment.

Materials: toy building materials such as blocks, Legos, Lincoln Logs, and Tinker Toys

Procedure: Place the materials in various locations around the room and on the floor. Divide the students into groups and assign them to locations. Provide time for them to construct whatever they want with their materials.

Closure: Have the students leave their constructions and move about the room to see what others have made. Have them take all their constructions apart before leaving the room so the materials will be ready for the next session.

Science Discovery Day

Activity Ideas (cont.)

Cookie Geology

Goal: Discover the differences between rocks and minerals.

Materials: cookies with many solid ingredients such as chips, nuts, raisins; samples of rocks and minerals; paper towels

Procedure: Display the rocks and minerals, letting students observe them. Give each student a cookie and a paper towel. Tell them to break the cookies in half, crumbling one half. They can then make piles of the solid ingredients (raisins, etc.) in the crumbled half.

Closure: Students compare their cookie contents. Explain that the cookie represents a rock, the ingredients they piled represent the minerals, and the remaining solid half is the matrix or leftover rock. Show them the mineral and rock specimens. Explain that each mineral is separate from the others, but when they are mixed together and held by the matrix, they make a rock.

Create an Insect

Goal: Make a simulated insect, using various materials to represent its body parts.

Materials: items which can be used to simulate insect legs, body sections, and antenna such as pipe cleaners, egg cartons, cotton tips, cotton balls, toothpicks, colored paper, straight pins, clear tape, clay; pictures of winged and non-winged insects which clearly show their body parts; live insects in plastic bags or jars; magnifiers; overhead projector

Procedure: Show the pictures and live specimens of insects. Examine the latter on the overhead projector as well as with magnifiers. Discuss the parts of an insect. Tell the students they are to use the materials to create their own insect; however, their imaginary insects should be realistic and include common insect components (antennae, thorax, etc.) Name the insects. They should name their insects on the basis of one of its characteristics.

Closure: Have each student show his or her insect and tell its name. Let them compare these with the pictures and specimens. They can keep their models as souvenirs of this activity.

Fingerprinting

Goal: Learn about fingerprint patterns and compare fingerprints.

Materials: soft lead pencils, scratch paper, white paper, hand lenses, overhead projector, transparency of fingerprint examples

Preparation: Make a drawing of one hand on a piece of white paper. Prepare a coating for making fingerprints by rubbing the pencil on scratch paper and pressing the finger against the patch of pencil lead. Press the finger to the white paper within the outline of that finger. Repeat this with all the fingers and thumb of the hand. Make a transparency of this.

Procedure: Show the transparency to the students. Let them examine the patterns they see. Explain that every fingerprint is different and that they are going to make their own fingerprints to compare them. Distribute a hand lens, pencil, scratch paper, and white paper to each student. Have them duplicate your preparation procedures with each finger.

Closure: Have students examine their prints with a hand lens and then exchange their prints to compare them. Students should take their prints with them as a souvenir of this activity.

Science Discovery Day

Activity Ideas (cont.)

Good Vibrations

Goal: Make a musical instrument from a variety of materials.

Materials: items which can be turned into musical instruments—large balloons, juice cans with both ends removed, glass bottles or test tubes, water, rubber bands, shoeboxes, waxed paper, and combs; wind, string, and percussion musical instruments such as a drum, kalimba (thumb piano), guitar, flute

Procedure: Do a demonstration to show the students that sound is really vibrations that travel to our ears. Have the students sit at a long table and listen to you knock on the table. Tell them to look away from you and press their ears to the table. Repeat the knocking on the table. Ask them which sounded louder (the knock on the table does). Explain that when you knocked on the table the second time, the sound made both the air and table vibrate. The sound traveled through the air to their ears slowly the first time since the air molecules are spread far apart. It traveled faster through the solid table, which has tightly packed molecules that let vibrations travel faster.

Show the musical instruments and demonstrate how they work. Explain that every musical instrument depends upon vibrations to make sounds we hear. Tell the students they will use the materials provided to construct musical instruments. Let them work in small groups to make their instruments and to test them.

Closure: Have the groups show their musical instruments and demonstrate how they work. Compare these to the instruments shown at the beginning.

Mirror Magic

Goal: Investigate the result of viewing images through mirrors.

Materials: flat, rimless hand mirrors; large print, uppercase letters on a page; colorful photos from magazines; portrait picture of a full face

Procedure: Distribute a mirror and a sheet of letters to each student. Have them place the mirror over a letter, perpendicular to the paper, to see what the letters look like. Have them turn the mirror on the letter to discover how many different letters they can create with their mirror. Let them try this with all the letters to see how many different letters they can make.

Distribute the colored magazine pictures and let students try to change the images with their mirrors. Have them pair up and use two mirrors at right angles to each other over the pictures. Tell them to open and close the mirrors like a book and watch in the mirror to see how the images of the pictures change into multiple images. Let them hold the two mirrors at right angles and slowly pull them down the page with the colored pictures.

Closure: Distribute the face photographs. Have the students place the mirrors on the faces so they are cut in half. Have them look at the new faces this creates. Let them turn the mirror around and then look at the other half of the face to see another view.

Science Discovery Day

Activity Ideas (cont.)

Mealworms

Goal: Investigate the characteristics of mealworms.

Materials: jumbo live mealworms (available in pet or fishing stores), hand lenses, file cards, pencils, metric rulers, small sticky notes

Procedure: Distribute the materials to each student. Let them observe their worms for a few minutes. Have them draw their mealworms on their cards. Use the hand lens to count the number of segments the mealworm has. Record these on the board. Measure the worm in millimeters and record the length on the note.

Closure: Make a bar graph on the board, showing the lengths of the mealworms. (Use the sticky notes.) Have students discuss what they learned about their mealworms.

Parachutes

Goal: Create a paper parachute and get it to carry a passenger safely to the ground.

Materials: large sheets of colored tissue paper, self-adhesive hole reinforcements, string, colored pens, clip wooden clothespins, pictures of parachutes (*optional*: a real parachute)

Procedure: Show the pictures of various types of parachutes. Tell students they will each make one. Distribute the tissue paper and hole reinforcements to each student. Demonstrate how they should put reinforcements at the four corners of the tissue paper and halfway between each corner. Poke a hole through the center of each reinforcement. Tell them to decorate and put their names on the parachutes, being careful not to tear the tissue. Have each student paint a face on one end of his or her clothespin. They should then tie 10" (25 cm) of string through each hole and then gather all ends and tie these in a knot. Clip the clothespin to the knot.

Closure: Take the parachutes outside, away from obstructions. Have students stand apart from each other, holding their clothespin persons in one hand, and then throw the persons high into the air. They should soar slowly to the ground as the parachutes open. Repeat to perfect the flight.

Scavenger Hunt

(This activity may be conducted by older students as well.)

Goal: Locate things of science found around the school.

Materials: list of science things, such as something green which is alive, four-leaf clover, insect, and colored rock; container to hold materials, including bug box

Procedure: Look over the grounds before the event and compose a list of about ten items for students to collect. Divide the students into pairs or groups of three. Give each group a list and container. Review the items. Arrange a signal at which they will return (bell, whistle) and then let them search. Older students and the presider can oversee the groups without helping to search.

Closure: Call all the students together again. Have the groups show what they found. Scatter the items they found around the grounds again before the next session.

Science Discovery Day

Activity Ideas (cont.)

❏ Grades 5–6

Building a Flashlight

Goal: Create electric circuits with batteries and small bulbs.

Materials: D-cell batteries, thin wire (telephone wire will do), flashlight bulbs, toilet paper rolls, masking tape, aluminum foil, scissors

Preparation: Cut the wire into 12" (30 cm) lengths and strip the insulation from both ends.

Procedure: Distribute a battery, bulb, and one wire to each student and have them put these together to turn on the bulbs. It will take some time until the students accomplish this task. The drawing below shows four ways of connecting these parts to light the bulb.

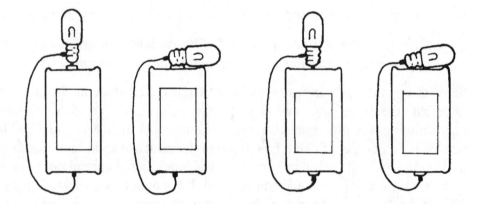

Once students have succeeded in discovering what is needed to light the bulbs, divide them into pairs. Distribute a roll of tape, piece of aluminum foil, scissors, and toilet paper roll. Challenge them to make a flashlight using two batteries but only one bulb and one wire. The batteries must be aligned positive to negative in the tube. One method of making the flashlight, complete with a switch, is shown below.

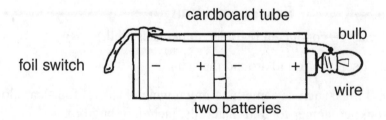

Explain that they need to be able to turn their lights on and off and that they should project the light in the dark.

Closure: Have students share what their flashlights look like and what they can do (by darkening the room).

Activity Ideas (cont.)

Fish Dissection

Goal: Learn about the anatomy of a fish.

Materials: 12" (30 cm) fish (mackerel are good and may be purchased at many bait stores), dissecting scissors, picture of the exterior and interior of a typical fish (from an encyclopedia), tweezers, newspaper, paper towels, hand lenses

Procedure: This activity is best done outside. Divide the students into pairs and distribute the materials. Have them examine the fish to locate the parts shown on the encyclopedia drawing. Tell them to use tweezers to pull open the gill cover. Next, they should open the mouth to look inside and through the gills. Have them pull off a scale and examine it with the lens to see the growth rings. Count and examine the various fins. Discuss their purpose in helping the fish swim.

Distribute the pictures of the fish interior and review the organs and their locations. Have students use scissors to cut out one section of the gills and examine it to see how feathery it is. Let them cut out the eye, being careful to cut deeply enough to remove it and to see the optic nerve at the back. They can cut the eye open and find the lens, which is spherical in shape.

The fish should be cut open from anus to mouth along the stomach and then perpendicular to the lateral line. The cut continues along the lateral line and then down to the anus. This skin should be removed carefully to reveal the internal organs. Have the students find the heart and liver by matching what they see inside their fish to the drawing. Tell them to remove the heart, which is triangular in shape. If large enough, it can be cut open to view its two chambers. Point out how close it is to the gills and that there are no lungs. Fish pump blood through their gills; the gills act like lungs, exchanging carbon dioxide for oxygen extracted from water. Have them continue to look for body parts, such as the stomach and intestines. They may open the stomach to see if the last meal is still there.

Closure: Students can carefully cut open the back of the skull to find the small brain. Have them discuss the similarities and differences between the anatomies of fish and humans.

Flower Dissection

Goal: Discover the parts of a flower.

Materials: whole flowers (tulips and lilies work well), variety of other flowers, illustration of cutaway view of a flower, magnifiers

Preparation Check an encyclopedia or other resource to find a drawing of the cutaway view of a flower and make copies of it. A florist shop may be willing to give you free lilies or other large flowers which show both male and female parts. Additional flowers may be available from gardens, or you can use wildflowers.

Science Discovery Day

Activity Ideas (cont.)

Flower Dissection (cont.)

Procedure: Give each student a flower and a cutaway illustration. Have them find the stem, petals, sepals (not on all flowers), stamen (male part with pollen), and pistil (female part with sticky stigma on top). Tell them to begin dissecting the flower by carefully pulling off the petals. They should then remove one of the stamens and look at the top of it (anther) with the magnifier to see the pollen. If they run their finger over the anther, some pollen may stick to it. They should carefully remove all stamens to expose the pistil. Tell the students to feel the top of it to see if it is sticky. Explain that pollen is carried to the stigma by wind, insects, or other means and sticks there. It begins to grow a long tube down the pistil to the ovary. Let the students slit open the ovary and look with their magnifiers. They may see seeds which have begun to form there.

Closure: Have students open other types of flowers to compare. Some of these may be multiple flowers (e.g., dandelions) while others may have only male or female parts.

Magnet Pictures

Goal: Make a permanent image of the magnetic field of various magnets.

Materials: different kinds of magnets, iron filings, sun-sensitive paper, paper plates, old nylon stockings, rubber bands, small paper cups, basin of water, newspaper, heavy books

Preparation: Cut the stocking into pieces to fit over the cups. Place three tablespoons (45 mL) iron filings into the cups. Cap with nylon stretched across the top and secured with a rubber band.

Procedure: This activity requires sunlight to expose the sun sensitive paper. Demonstrate how to show the magnetic field with iron filings by placing a magnet on one paper plate and covering it with another. Sprinkle filings through the nylon cover over the top plate in the area of the magnet beneath it. The filings will be drawn into the magnetic field and concentrate at the poles. Arcs will form between the poles. Only a fine mist of filings is needed to see the magnetic field.

Distribute magnets, two plates, and a cup of filings to each student. Let them experiment with outlining the magnetic field. The filings can be returned to the cup after each pattern is finished.

Have students make an arrangement of their magnets on one plate and cover it with the other. Give each student a piece of sun sensitive paper to label with their names. Have them place the paper face up over the top plate. The filings are then sprinkled over the paper to show the magnetic field. All materials are carefully carried into the sunlight and left to sit until the color fades. If the sun is bright, this may be no more than a few minutes. Return the materials to the classroom. Pour the filings onto the plate and put the paper in water to develop. After the image begins to appear, place the paper between layers of newspaper cover it with a book and let it dry.

Closure: Have the students compare their magnetic field designs. If two magnets were used, can they see the repelling or attracting forces which flowed between them?

Science Discovery Day

Activity Ideas (cont.)

Making a Compass

Goal: Make a simple compass.

Materials: small magnets (available at Radio Shack), strong bar magnet, steel straight pins, chips of foam cup or meat tray, aluminum muffin liners or pie pans, water, compass, globe with north magnetic pole marked (northern Canada area)

Preparation: Use thread to hang the bar magnet from the ceiling so it will be balanced and yet swing freely. Tap one end to set it spinning and watch to see where it points when it stops. Do this several times. The magnet should always point in the same direction when it stops, which is north-south. Mark the magnet's north end with a piece of masking tape. It is now a compass.

Procedure: Have students look at the hanging bar magnet and watch to see what happens when you tap it. Do this three times so they will see it always points in the same direction when it comes to rest. Show the students the compass and demonstrate how the floating needle always points north. Tell the students that a compass needle is really a magnet, like that hanging from the ceiling, and is attracted to the magnetic poles of the earth. Explain that the north magnetic pole does not line up with the north pole of the axis which points to the north star but is located close to that spot in an area of northern Canada. Show them where this is on the globe. Have students find which direction is north relative to the classroom. Mark this on the wall.

Tell them they are going to make a simple compass. Distribute a magnet and steel pin to each student. Show them how to rub the pin across the magnet in one direction, about fifty strokes, pressing the pin against the magnet as they rub. This will magnetize the pin. The pin is now threaded through the foam chip so it can be placed on the water in the aluminum container and float. When the pin comes to rest, it should point north-south, as the hanging magnet is doing. If it floats to the side of the container, it should be gently pushed into the center.

Closure: Have the students tap the chip to send the pin spinning gently. Tell them to watch to be sure it always points in the north-south direction. They may keep their compasses.

You Are Cleared for Takeoff

Goal: Build a simple, working model of a plane.

Materials: inexpensive balsa wood planes with rubber band, wind-up propeller, standard-size white paper, scissors, masking tape, clear tape, paper clips, file cards

Procedure: The presenter should explain how planes are able to fly. Students make a paper airplane, using the white paper. This can either be a plane of their own design or one the presenter demonstrates. They fly these inside the room and then use tape, cards, and paper clips to modify and improve them. These are tested again to find those which fly longest, fly the greatest distance, and do rolls. Students construct the balsa planes and take them to an open field to fly. They modify the planes with additional flaps and twists of the propeller to improve their flights.

Closure: Discuss the flights and relate them to the initial information about how planes fly.

Science Discovery Day

Activity Ideas (cont.)

You Are Go for Launch

Goal: Experience a simulated launch and flight to the moon.

Materials: coveralls from a paint or hardware store, space mission patches, helium filled balloon, large box, copy of script from *Space: Intermediate* (see Resources, page 95), slides showing moon and Earth from space (optional; see Resources, page 96)

Procedure: Students take the roles of crew members and passengers on a simulated flight from Earth to the moon. Enroute they deploy a satellite which will orbit the moon. The script provides information which is based, as much as possible, on current scientific information for such a journey.

Closure: Have students discuss the flight and how they feel about going to the moon.

❑ **Grades K–6**

Science Mural

Goal: Create a mural which depicts things related to science.

Materials: large sheet of butcher paper, white drawing paper, crayons or colored pens, pencils, rulers, compasses, restickable glue, scissors, science magazines

Procedure: Students are asked to make drawings of things which represent science, such as a flower, rainbow, and spaceship. Let them look through science magazines to get ideas for their pictures and then draw them on white paper. These should be colored, cut out, and temporarily mounted on the large paper. After the final drawings are made in the third session, all pictures should be arranged on the large paper to create a science mural.

Closure: Print a title on the mural and mount it in an area where all students can view it at the end of the day.

Science Discover Day Challenges

Guest Presenters

The task of finding 40 presenters may seem daunting. However, there are many science experts and hobbyists in your local community, and several are likely to enjoy taking part in such an important and exciting day. Here is a list of some of those who may be available to offer unique presentations. Send them invitations and present forms and follow up with telephone calls.

Hospital personnel: doctors, nurses, nutrition experts

Non-teaching school personnel: school nurse, janitor, gardener

Humane Society: animal care technicians and experts

Museums: curators, technicians (Many museums have educational outreach programs.)

Colleges and universities: science professors, student teachers, graduate students

Retired teachers groups: wide variety of skills and experience available here

Police department: criminologists, forensics experts

Fire department: fire equipment and safety experts, paramedics

Parents: a wealth of information, experience, and know-how to be found here

Other Challenges

There is always the possibility of the unexpected which can alter plans for an event of this magnitude. These may include bad weather, an absent presenter, new students enrolling the day of the event, equipment failure, and last-minute equipment requests by presenters. Try to anticipate some of these possibilities in advance by making plans for what to do if something happens. If it rains, outside activities will need to be relocated indoors or may need to be cancelled and the students reassigned to other activities. If a presenter is unable to come, one of the school staff members may need to step in with an alternate activity. Prepare two or three alternate activities that are easy to organize and can be offered to all grade levels. Be sure to appoint someone who will be in charge of handling the equipment and material requests during the event.

Science Discovery Day will be an outstanding event if there has been good planning and many people are involved to assure its success. Students, staff, and guests will find this to be an exciting day, thus making the efforts required in arranging the event more than worthwhile. Everyone will be eager to repeat this experience another year.

Inventors Convention

Overview

Inventors Convention is designed to provide opportunities for elementary students to learn about inventions and then to create their own. The goal is to invent something useful that has never before been made. The following lessons develop student background in invention history.

- **Invention Detectives:** Students locate inventions in the classroom and at home. How they are used is described. The raw materials used in their manufacture are listed.

- **History and Evolution of Inventions:** There are several components to this section of study.

  First, a brief overview of inventions beginning in prehistory is provided for the students to show why inventing is so important. Next, antique inventions are examined and discussed. These are compared to versions of the invention which still may be in use today. Finally, student teams research the evolution of certain inventions. They create a presentation of their information, including a time line of the evolution to show how it changed into what it is today.

- **Famous Inventors:** Students gather information about famous inventors and their inventions.

Once students have developed their understanding of inventions, the next step is for them to create one. They will do a variety of activities leading to their final product.

- **Useful Inventions:** Using ordinary materials, students attempt to turn them into a musical instrument, a tool, or a toy.
- **The Need to Invent:** Students create an invention which could help the physically challenged person.
- **Being an Inventor:** Students make plans and drawings for their own inventions. They construct working models of them.

Funds may be needed for materials used in building the inventions. If the school supply budget does not cover the costs, request help from the PTA. Contact local industries and business organizations for financial support or contributions of materials. Parents and people from these organizations may also be willing to work as mentors to the young inventors. Linking students with mentors from the community could introduce students to careers which they might not otherwise have known about.

The activities in this chapter are designed for grades 3–6. However, primary students can also be involved in the excitement of inventing. This can be done by using activities appropriate for primary students or modifying the activities so they are suitable for their skill levels.

The grand finale is the Inventors Convention where all students display and demonstrate their inventions. This convention may involve one or more classes or everyone in the school, including staff. Parents and other visitors should be invited to attend the Inventors Convention. Suggestions for organizing such an event are provided on pages 66 and 67.

Inventors Convention

Invention Detective

Overview: Students will search for inventions in the classroom and at home.

Materials:
- *Invention Detective* work sheet (page 44, one per student)
- transparency of Invention Detective work sheet
- transparency marking pen
- parent letter (page 45, one per student)

Motivator:
- Discuss the meaning of the word invention so students realize it is something which uses natural materials to create other things we can use (e.g., shoes from leather).
- Explain that inventions are a creation of something which never before existed.
- Ask the students to look around the room to locate at least three different inventions.

Procedure:
- Distribute the *Invention Detective* work sheet and use the transparency of this form to complete some examples of inventions students found in the room.
- Instruct students to write several of the items they found around the room on the form. Let them complete the information about these inventions.

Homework Assignment:
- Attach the parent letter to the students' *Invention Detective* lists. Tell them to take their forms home and complete them by finding more inventions there. Discuss where they might look for unique inventions, such as in the garage or refrigerator.
- Read the letter to the students so they know the list is to be returned for the next class and that they are urged to bring antique inventions (or their pictures) for use during their study.

Closure:
- Share the items on the students' *Invention Detective* lists.
- Discuss how these inventions depend upon natural materials and people's imaginations for using them in unique ways to make useful items.
- Explain that they will be using the unique inventions in their next class.

Inventors Convention

Invention Detective (cont.)

You are an *Invention Detective* searching for as many inventions as you can find. As you discover them, list them here and give as much information about them as you can.

Name of Invention	It is used for	It is made of

Inventors Convention

Parent Letter for Invention Detective Assignment

Date: _____

Dear Parents,

Your child has begun a study of inventions. The students started by finding and describing inventions in our classroom. The next step is to find inventions they use at home. These are to be added to the attached Invention Detective form and returned to school tomorrow. Your help would be appreciated in this assignment. Please help your child find some unique inventions which you may have in your home, garage, or yard. Help your child describe its use and, if possible, list the raw materials from which it was made. You will find examples of how this is to be done by looking at the three inventions from the classroom which are described on the chart.

Be sure to ask your child to tell you about the class's invention study. Eventually, the students will be making inventions of their own, complete with working models. You will be invited to our Invention Convention at the end of our study.

Thank you for your help.

Cordially

_____

Inventors Convention •••

History and Evolution of Inventions

Overview: Students will learn about the history and evolution of a variety of inventions through classroom study, homework, and team research. These activities should be spread over time (two to four weeks) for students to gather their data and to prepare their presentations. (**Note:** The lessons have been divided into two sections, Part I and Part II, for the sake of clarity and so as not to overwhelm either the teacher or students.)

Part I

Materials:

- Parent Letter for History of Inventions (page 48, one per student)
- examples of inventions which have changed over the years but are still used today
- antique inventions or pictures of them (some of which are now obsolete)
- Important Inventions in History (pages 49 and 50, one per student)
- transparency of Important Inventions in History
- Evolution of the Lollipop (page 51, one per student)
- transparency of Evolution of the Lollipop
- transparency marking pen

Procedure:

- At least a week before beginning this lesson, send home a letter to the parents (page 48) which asks them to share with the students any antique inventions they might have.
- Show the students the examples of antique inventions and discuss them. Wherever possible, compare early inventions with those we use today, such as a compact disc player rather than a record player (including early Victrolas and 33-rpm turntables). Discuss the transitions of these inventions.
- Discuss inventions which did not last long, such as the 8-track tape player or beta video machine.
- Show the transparency of Important Inventions in History and briefly discuss this so students understand the progression of inventions.
- If available, invite an antique collector to display and explain some early inventions.

Closure:

- Use Evolution of the Lollipop (page 51) to show how an invention can evolve.
- Surprise the students by serving lollipops after the discussion.

Homework Assignment:

- Send home a copy of Evolution of the Lollipop (page 51). Instruct students to design a new form of the lollipop with the help of their families. This can be done as a design on paper or as an actual lollipop. Give them several days to work on this assignment. Have them bring their new lollipop inventions to share with their classmates. (See page 47.)

··· *Inventors Convention*

History and Evolution of Inventions (cont.)

―――――――――― Part II ――――――――――

Materials:

- lollipop inventions made by students (see page 46)
- History of Inventions work sheet (page 52, one per student)
- lined file cards
- reference materials (e.g., encyclopedias, books about inventions, Internet access)

Procedure:

- Let the students share their new lollipop inventions.
- Tell the students that the evolution of the lollipop is only one example of how inventions can change over time. Explain that they will be divided into teams to research the history of an invention and to present this information to the class.
- Divide the students into teams of two or three students. Distribute copies of page 52 (preferably one per student, but one per team is also acceptable).
- A list of topics is provided below. Select those which would be appropriate for your students, adding any others you would like. Write these on the board and discuss them briefly. Let each team choose its topic for research. If you wish, allow the teams to propose topics of their own.

> ❏ airplane   ❏ automobile   ❏ computer   ❏ musical instruments
> ❏ radio   ❏ rockets   ❏ roller skates ❏ ships
> ❏ television ❏ typewriter

- Show the resources which have been collected for them to use as they research their topics. Distribute file cards to each team and review the instructions given on the work sheet (page 52) that show how to use the cards. (Alter these instructions to meet the level of your students.)
- Permit time for students to work on this project during the school day, as well as at home. Encourage them to share resource ideas with other teams.
- Encourage students to be creative in the way they develop their presentations.
- Instruct students to include a time line of the dates for each stage in the invention's development. They should also display pictures or drawings of these stages as well as illustrations of the invention's evolution. Finally, they should include a visual prediction of what the invention might look like in the future.
- Provide bulletin board space for each team to display its time line and pictures.

Closure:

- Schedule a presentation time for each team.
- Schedule a time for other classes and parents to visit the bulletin board displays.

Inventors Convention

Parent Letter for History of Inventions

Date: _____

Dear Parents,

As a part of our study of inventions, we are going to investigate antique inventions which are no longer in use as well as inventions which were used in the past but have evolved into different forms today. An example of an antique invention no longer used today is a shoe-button hook. The iron is an invention which evolved and is still in use. When first invented, it had to be heated on a stove, but it has evolved into today's electric steam iron.

I would like the students to see as many examples as possible of these types of inventions, and I would appreciate your help. If you have any inventions you would be willing to loan us, please bring them to school. Be sure to label them with your name. These will be well cared for and returned to you in about a week. Another option would be to bring the item(s) to class and to tell the students about it yourself. If you prefer this option, please contact me, and we can discuss a convenient scheduling time for your visit.

Thank you for your help.

Cordially,

_____

Inventors Convention

Important Inventions in History

Inventions began with early man, but most of the important inventions were developed during the last 600 years. These inventions changed our lives in every way, including agriculture, communications, industry, science, transportation, and warfare. Inventions continue even more rapidly today; there appears to be no end in sight as to what can be invented.

A brief overview of major inventions is provided here so that students may begin to understand how important it is to continue inventing. Make transparencies of this page and page 50 to show the students. Explain some of the inventions which may not be familiar to them. Discuss how these inventions have helped change our lives.

flint tools about 1,750,000 B.C.	plow 5000–3000 B.C.	wheel circa 3500 B.C.	Archimedean screw 200s B.C.
paper circa 100 B.C.	magnetic compass 1100s	cannon circa 1350	printing from movable type circa 1440
compound microscope circa 1590	telescope 1608	steam engine 1690–1769	spinning jenny circa 1764
balloon 1783	steamboat 1787–1807	cotton gin 1793	food canning 1795–1809

© Teacher Created Materials, Inc. #2509 *Beyond the Science Fair*

Inventors Convention

Important Inventions in History (cont.)

steam locomotive 1804	stethoscope 1816	Portland cement 1824	photography 1826
safety pin 1849	safety elevator 1853	hypodermic syringe 1853	Bessemer steel-making circa 1860
internal-combustion engine 1860	dynamite 1867	typewriter 1867	barbed wire 1873
electric motor 1873	telephone 1876	phonograph 1877	incandescent light 1879
skyscraper 1885	gasoline automobile 1885	zipper 1893	motion picture mid-1890s
radio 1895	x-ray machine 1895	safety razor 1901	airplane 1903

#2509 Beyond the Science Fair © Teacher Created Materials, Inc.

Inventors Convention

Evolution of the Lollipop

1850s The lollipop is born . . .
The lollipop is invented when someone puts small balls of sugar candy on the end of a pencil.

1880s   . . . and improved . . .
A candy merchant in Connecticut puts chocolate caramel taffy on a stick, claiming it is easier to eat that way.

1892 . . . given a name . . .
George Smith, a candymaker in Connecticut, affixes hard candy to a stick and calls it "Lollipop" after a popular racehorse of the day.

1924 . . . and a new form . . .
An American classic, the Dum Dum Pop, is first marketed.

1931 . . . and then made even tastier . . .
The Tootsie Pop, a lollipop with filling, is created.

1993 . . . with a technology twist . . .
The first interactive lollipop, the Spin Pop, is introduced.

1998 . . . and influenced by television.
A new lollipop, the Sound Bite, is created to keep up with the demands of modern technology. It plays music in the eater's mouth.

Now it is your turn! What does the future hold for this favorite treat?

Inventors Convention

History of Inventions

Your team has been given the assignment to find out all you can about the history of an invention. Follow the suggestions below to guide you as you collect information.

1. **Use lined file cards to write your information,** as shown in the example below. Make a new set of file cards for each resource.

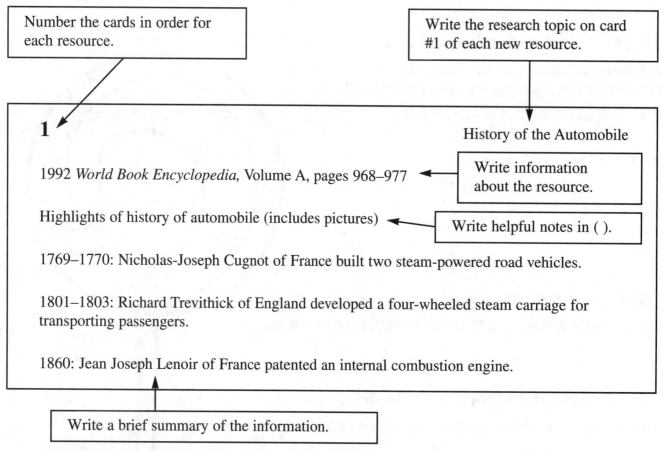

2. **Use the note cards to prepare a report to the class.**
    - Write the dates on the unlined side of individual file cards.
    - Place the date cards in order on a bulletin board to form a time line.
    - Place pictures and brief information on the bulletin board beside the dates.

3. **Make a drawing of how this invention may look in the future.** Place the picture and a date card for it on the bulletin board.

4. **Think of an interesting way to present your information to the class.** Some suggestions are listed below.
    - Use the History Channel format and present the information about the invention as a television program.
    - Have team members play the roles of the different inventors. Each will tell about his or her invention in the order in which they were built.
    - Work as a team to be a living model of the invention, pantomiming its movement and purpose. Each person will be a piece of the collective whole.

Inventors Convention

Famous Inventors

Overview: The biographies of famous inventors will be studied to discover how they developed their inventions.

Materials:

- resource materials about inventors
- optional: Internet access
- file cards

> 1
> 1992 *World Book Encyclopedia*, Volume A, pages 968–977 — History of the Automobile
> Highlights of history of automobile (includes pictures)
> 1769–1770: Nicholas-Joseph Cugnot of France built two steam-powered road vehicles.
> 1801–1803: Richard Trevithick of England developed a four-wheeled steam carriage for transporting passengers.
> 1860: Jean Joseph Lenoir of France patented an internal combustion engine.

Procedure:

- Review what students have learned thus far about inventions.
- Explain that they will now learn about some inventors who created things to improve our lives. Tell them that each student will research an inventor and make a presentation to the class. Provide any guidelines for the presentation that you desire. A partial list of inventors whom students may research is provided below. Select those who are appropriate for your students, adding any others you would like. (Students may also research inventors as they prepare their reports for the History of Inventions activity, page 52.)

American
Bell, Alexander Graham
Carver, George Washington
Curtiss, Glenn
Deere, John
Eastman, George
Edgerton, Harold
Edison, Thomas A.
Ford, Henry
Franklin, Benjamin
Fulton, Robert
Goodyear, Charles
Jefferson, Thomas
McCormick, Cyrus Hall
Morse, Samuel
Otis, Elisha
Pullman, George
Singer, Isaac
Steinmetz, Charles
Westinghouse, George
Whitney, Eli
Wright brothers

British
Bessemer, Sir Henry
Cartwright, Edmund
Dunlop, John
Hargreaves, James
Kelvin, Lord
Watson-Watt, Sir Robert
Watt, James

French
Cousteau, Jacques-Yves
Daguerre, Louis

Jacquard, Joseph
Lenoir, Jean J. E.
Montgolfier brothers

German
Benz, Karl
Bunsen, Robert
Daimler, Gottlieb
Diesel, Rudolf
Gutenberg, Johannes
Siemens, Ernst

Other Inventors
Archimedes
Da Vinci, Leonardo
Marconi, Guglielmo
Nobel, Alfred

Closure:

- Schedule student presentations so they will be done in chronological order and grouped by the types of inventions created by each inventor.
- Have the students present their material to the class and parents, if desired.

© Teacher Created Materials, Inc.   #2509 Beyond the Science Fair

Inventors Convention •

Useful Invention

Overview: Using ordinary materials, students attempt to turn them into inventions which are useful in solving a problem.

Materials:

- invention work sheets (pages 55–57, one per group)
- items needed, as listed on the work sheets
- scissors, tape, string, glue, stapler, water, and other items which teams may need during the construction of their inventions

Preparation:

- Make sets of materials for students to use in creating an invention as shown on the work sheets. The sets will be duplicated so that two teams will work with the same set of materials to see if they can use them in a different way.
- Prepare a table containing items all groups may borrow during their work.

Procedure:

- Briefly discuss what students have learned about inventing. Be sure they know that inventions often come about because there is a problem which needs to be solved. Also, inventors need to be creative and look at using materials in different ways than most people would in order to turn them into new inventions.
- Explain that the students will be divided into teams. Tell them that each team will be challenged to use a set of materials for inventing a specific type of item. Let them know that each invention task will be done by two different teams. The challenge will be that each team must make a different invention from the same set of materials.
- Divide the students into six teams and distribute the work sheets. Let each team read the instructions. Be sure the pairs of teams working on the same task are separated in the room and do not exchange ideas.
- Show the students where in the room the general team materials are located.
- Distribute the sets of materials to the teams. Instruct them to examine and discuss these before they begin the task. Without giving them any ideas, monitor their progress and offer them encouragement to be creative.
- This lesson may require more than one class session to complete in order to encourage the students to apply their creative skills to the task.

Closure:

- Have students show their inventions and demonstrate how they work.
- Place the inventions on display in the classroom, along with the work sheet for each team.

Inventors Convention

Invent a Musical Instrument

You are a team of musicians which wants to enter a contest. It requires that a new type of musical instrument be made with the materials listed here. You must be able to play a simple tune on the instrument.

Materials:

- 3 shoeboxes
- 10 tongue depressors
- 6 glass bottles
- rubber bands of different widths
- 12 drinking straws
- large metal nail

Getting Started:

- Look at the materials you have available to construct the musical instrument. Try using the materials to see if you can make sounds with them.
- The team needs to decide which of these materials will be used to make your musical instrument. (Remember, you must be able to play a song on the invention you make!)
- In the box below, draw a design for your musical instrument. Label the materials that will be used in it and show how your instrument will be played. (You are welcome to use the general materials provided, such as scissors and tape, to construct your instrument.)
- Assemble your musical instrument and play a simple tune on it.
- Make any changes which may be needed to improve the instrument so it makes good sounds.
- Give your invention a name and write it in the design box.

Invention name: _____

Invented by: _____

Inventors Convention

Invent a Tool

You are a team of engineers. You have been given the materials listed below. Your challenge is to use them to make a useful tool. You will decide what the tool should look like and what work it will do for you.

Materials:

- 3 feet (90 cm) string
- box of large paper clips
- 6 tongue depressors
- 10 rubber bands
- 12 drinking straws
- 6 metal washers about the size of a nickel
- circle magnet about the size of a nickel and with a hole in its center

Getting Started:

- Look at the materials you have to use in constructing the tool. Try using the materials to see if they can be used to do some type of work.
- The team needs to decide which of these materials will be used to make your tool. Remember, you must be able to do some type of work with this invention.
- In the box below, draw a design for your tool. Label the materials that will be used and show how it works. (You are welcome to use the general materials provided, such as scissors and tape, to construct your tool.)
- Assemble your tool and test it to be sure it works.
- Make any changes which may be needed to improve the invention to make it work better.
- Give your invention a name and write it in the design box below.

Invention name: _____

Invented by: _____

•• *Inventors Convention*

Invent a Toy

You are a team of engineers. You have been given the materials listed below. Your challenge is to use them to make a new toy. This toy can be for a child of any age.

Materials:
- 3 unsharpened pencil
- box of small paper clips
- 6 drinking straws
- 2 feet of heavy string
- 6 metal washers about the size of a nickel
- 5 circle magnets about the size of a nickel and with holes in the center

Getting Started:
- Look at the materials you have. Try using the materials to see if you can get any ideas how they may be used to make a toy.
- The team needs to decide which of these materials will be used to make the toy. Remember, the toy can be used by a child of any age. Toys for young children must be simple but also safe so they cannot be swallowed.
- In the box below, draw a design for the toy. Label the materials to be used and show how it works. (You are welcome to use the general materials provided such as scissors and tape to construct your toy.)
- Assemble your toy and test it to be sure it works.
- Make any changes which may be needed to improve the invention to make it work better.
- Give your invention a name and write it in the design box below.

Invention name: _____

Invented by: _____

Inventors Convention •

The Need to Invent

Overview: Students experience what it is like to be physically challenged and create an invention that will help them.

Materials:

- copies of page 59 (one per student)
- copies of pages 60 and 61 (enough for all students when divided into pairs)
- one-inch (2.5 cm) wide, removable, adhesive bandage
- soft earplugs used for swimming
- eye patch (black fabric or paper will do)
- baseball bat
- sleep masks
- index cards
- dark cellophane
- cups of applesauce
- drawing paper
- wheelchairs
- cloth strips (2" x 3' / 5 cm x 90 cm)
- wide stretch bandage with fastener
- spoons
- crutches
- ticking clock
- tennis balls
- disposable combs

Preparation:

- Send home the parent letter before doing this activity.
- Contact parents and the school health clerk or nurse to find crutches and wheelchairs which may be used for this activity. Sleep masks and earplugs are usually available in drug stores.

Procedure:

- Ask the students if they have ever had to use crutches or a wheelchair. Have them share what this was/is like and the problems they may have had. Explain to the students that inventions are often created to help people who are physically challenged. Discuss how these have helped them.
- Tell the students that they are going to be teamed with a partner and that each pair of students will be given a temporary physical challenge. Let them know that each member of the team will spend time experiencing the challenge. Explain that the purpose of this activity is for them to design an invention that will make life easier if they have to live with this challenge forever. They will not actually make the invention but draw it instead.
- Divide the students into pairs and let each team choose its physical challenge card from a bag. Have them read the information on the card and discuss what the rules will be during this time.
- Complete the challenge.

Closure:

- Distribute drawing paper to the students and have them design their inventions. They should label them and explain how they will be used.
- Have each team post its design on the bulletin board and explain it to the class.
- Compare the inventions which were made for the same type of physical challenge.

• Inventors Convention

Parent Letter for Physical Challenge Activity

Date: _____

Dear Parents,

Our class is continuing our study of inventions. The students are about to be involved in an activity which will have them assume a physical challenge. These include forgoing the use of their legs so they must be in a wheelchair, wearing a sleep mask over their eyes to simulate blindness, and wearing earplugs to simulate loss of hearing. The students will work in pairs so only one partner will be physically challenged at a time. They will exchange roles so both partners have the opportunity to see how it feels to experience this challenge.

Safety will always be considered throughout this activity, and students will be closely monitored. During the time they experience this physical challenge, they will perform everyday tasks such as playing ball, writing, and moving around the classroom or outside.

At the end of this lesson, students will be responsible for designing devices that would help them live as normally as possible in spite of the physical challenge. They will make drawings of their inventions and explain them to the class.

We will need to borrow a wheelchair and adjustable crutches during this activity. If you have either one of these items that you could loan us for a few days, please contact me at school.

Be sure to ask your child about this activity and discuss how it made him or her feel to have a temporary physical challenge.

Thank you for your help in this project. You are welcome to visit our class during this activity or come to see the pictures of the students' inventions which will be displayed in our classroom in a few days.

Cordially,

_____

Inventors Convention •

Physical Challenge Cards

Physical Challenge: loss of vision in both eyes
Materials: sleep mask, applesauce, spoon
Getting Started: Put the sleep mask on so that you see no light.
Tasks: Walk around the room and outside. Pick up objects and try to identify them. Eat the applesauce.

Physical Challenge: both arms broken
Materials: two cloth strips, comb, applesauce, spoon
Getting Started: Cross your arms in front of you and have your partner use the cloth strip to tie them together. The hands should be free to move at the wrists.
Tasks: Write, sit and then stand up; comb your hair; take off your shoes and put them on again; and eat the applesauce.

Physical Challenge: loss of vision in one eye
Materials: eye patch, removable tape, tennis ball
Getting Started: Tape the patch over one eye so you cannot see.
Tasks: Walk around the room and outside. Play catch with your partner.

Physical Challenge: both legs broken
Materials: strip of cloth, wheelchair, tennis ball
Getting Started: Sit in the wheelchair and have your partner tie your legs together with the cloth strip.
Tasks: Use the wheelchair to move around the classroom. Move to your desk chair. Go outside and move around the area. Play catch with your partner.

Physical Challenge: one arm broken
Materials: cloth strip, applesauce, spoon, comb
Getting Started: Have your partner tie the strip around the wrist of your writing hand and then around your chest. This will hold your arm against your body.
Tasks: Write, sit, and then stand up; comb your hair; take off your shoes and then put them on again; eat the applesauce.

Physical Challenge: cannot speak
Getting Started: You cannot speak, not even whisper.
Task: Try to tell your partner something without using your voice.

#2509 Beyond the Science Fair © Teacher Created Materials, Inc.

Physical Challenge Cards *(cont.)*

Inventors Convention

Physical Challenge: can only see straight ahead

Materials: two index cards, removable tape, tennis ball

Getting Started: Fold the cards in half and tape them on either side of your head so you cannot see around them.

Tasks: Toss the ball with your partner. Walk around the room and outside.

Physical Challenge: one leg broken

Materials: crutches, stretch bandage, tennis ball

Getting Started: Have your partner put the stretch bandage around your knee so it cannot be bent. Do not make it so tight that it cuts off the flow of blood. Remember not to put any weight on that leg.

Tasks: Walk around with the crutches. Sit and stand. Play ball with your partner.

Physical Challenge: lost use of one thumb

Materials: removable tape, comb, applesauce, spoon

Getting Started: Tape the thumb of your writing hand to the side of your hand.

Tasks: Write, pick up a spoon and eat the applesauce, and comb your hair.

Physical Challenge: partial hearing loss in both ears

Materials: earplugs, clicker (noisemaker)

Getting Started: Place the earplugs into your ears to cut out as much sound as possible.

Tasks: Listen to the sounds around you with and without the earplugs. Notice what sounds you can hear. Repeat this as your partner walks around you and uses the clicker.

Physical Challenge: born without thumbs

Materials: removable tape, baseball bat, tennis ball, applesauce, spoon

Getting Started: Have your partner tape both thumbs to your hands.

Tasks: Play baseball with your partner, write, and eat the applesauce.

Physical Challenge: limited vision

Materials: 3" x 6" (7.5 cm x 15 cm) strips of dark cellophane, same size cardboard strip, two 8" (20 cm) strings, tennis ball

Getting Started: Draw a rectangle on the cardboard, leaving a half-inch (1.25 cm) border around it. Cut this section out and tape the cellophane over it. Attach strings to the long ends. Tie the strip in front of the eyes.

Tasks: Walk around, write, and play ball.

Inventors Convention •••

Being an Inventor

Overview: Students will create inventions of their own.

Materials:

- Students will make lists of the materials they need for their inventions. Some of these can be supplied by the school, and others may be brought from home.
- transparency of Ideas for Inventions (page 63)
- copies of A New Invention (page 64, one per student)
- index cards

Preparation:

- Make a list of possible resources for funding which may be needed for materials to be used in making the inventions. They may include the PTA, local businesses, and civic clubs.
- Make a list of possible mentors to consult during the construction of the inventions. These may include parents, grandparents, and people from local businesses such as construction companies.
- Locate books which may give students ideas for inventions. (See Resources, pages 94–96.)

Motivator:

- Review the lollipops students invented (pages 46, 47, and 51) and their invention designs (page 58).
- Discuss how they got their ideas for these inventions.

Procedure:

- Instruct students to design another invention that will fulfill a need they or someone else may have. They will be expected to design the invention and then construct a working model of it.
- Show the transparency and discuss the ideas with the students.
- Let students select their teammates for this project or work alone. Distribute the work sheet and discuss the steps. Encourage them to research carefully for ideas before beginning their designs.
- Tell them that they may receive help on their designs and construction, if needed, but that they should be the ones who do most of the work rather than anyone who helps them.
- Establish for the students due dates which will provide sufficient time for them to do their best work in completing their invention for the Inventors Convention. Set target dates for the invention tasks and have the students record these on their work sheets.
- After students submit their invention ideas, check to be sure they can be constructed by the students. Suggest changes, if needed, for those that are too complicated or inappropriate.
- Give students index cards to list the materials they will need. These should be submitted with their designs. Gather materials through family donations or corporate support.
- During the construction period, allow time for students to try out various ways of building their inventions. Permit them to modify their designs as needed during the construction. This will take time but will develop their creativity and other skills. It will be best to permit class time for construction so that students do the work rather than someone at home. Mentors should serve as consultants rather than take the lead to build the invention.

Closure:

- Conduct an Inventors Convention. (See pages 66–67).

Ideas for Inventions

Students may need ideas to get started on an invention. You will find resources for this purpose on pages 94 and 95. A brainstorming session with the students may also be helpful. Let students suggest new inventions as you list them on the board. Here is a list to get you started.

New Products from Old Ideas

- airplane
- birdbath
- board game
- candy
- clothes hanger
- dessert
- hair clip
- paper clip
- playground equipment
- shoe
- ball
- bird feeder
- building blocks
- chair
- clothespin
- healthful snack
- musical instrument
- pencil
- pet house
- toy

New Uses for Old Products

- bedsheet
- drinking straws
- flashlight
- magnet
- paper towel roll
- plastic flowerpot
- rubber bands
- shoebox
- stool
- Styrofoam cup
- book
- facial tissue
- key ring
- milk carton
- plastic bottles
- newspaper
- ruler
- spoon
- stuffed animal
- toothpicks

Inventors Convention •••

A New Invention

Name(s) of the inventor(s): _____

	Project Deadlines
Target Date	**Task**
	Submit the invention idea for approval.
	Submit the design for the invention.
	Complete construction and testing of the invention.
	Complete a display for the Inventors Convention.

Getting an idea for your invention:
- Think of a problem that you might solve with an invention. This could be something which will be useful to you or to other people. Remember, you will need to build a working model of your invention so think of something simple. If your invention requires energy, use batteries and not electricity from a wall socket.
- Your invention may be an improvement on one which already exists or a mixing of one or more inventions to make a new one.
- Ask others what they need or could use to help you get ideas for something to invent.
- Read books which may have suggestions for simple inventions.

Submitting your invention for approval:
- Describe the invention you plan to make. Be sure to explain the invention in such a way that your teacher will understand what it is and what it will do. Give your invention a name.

Submitting your design for approval:
- Make a detailed drawing to show the invention.
- Label all the parts to show what materials you will use and what each part does.
- Make drawings to show several views of your invention, such as front, back, side, and top.
- Write a few sentences to describe how the invention will work and what it is expected to do.
- Use index cards to list all the materials you will need for construction. Make separate cards for materials you will bring from home and those you are asking your teacher to provide.
- Write the name of the invention and the inventors on the cards. Attach them to your invention drawing before giving it to the teacher for approval.

Completing construction and testing the invention:

This should be done without adult help, as much as possible. Be sure to ask for assistance if you need to use a dangerous tool such as a saw during the construction. Conduct tests on the model and change it as needed for it to work.

Completing a display for the Inventors Convention:

Follow the display instructions and do your best work so you will be proud of the results when others see your project.

Display Instructions

The display of your invention will be seen by everyone who attends the Inventors Convention. Follow the instructions below to tell about the invention, how you made it, and how it works.

1. Use a display board made of cardboard for your exhibit. This may be purchased at most office supply stores. It has three sides and will stand upright so that your invention can be set in the center of the area.

2. On separate papers neatly print the following:
   - name of the invention
   - name(s) of the inventor(s)
   - name(s) of those who helped you with this project
   - how you got the idea for the invention
   - how it works
   - what problem it solves

3. Display the drawing you made of the original design.

4. Show any photographs taken during the construction of your invention.

5. Be prepared to attend the Inventors Convention to demonstrate your invention and to answer questions about how it was built.

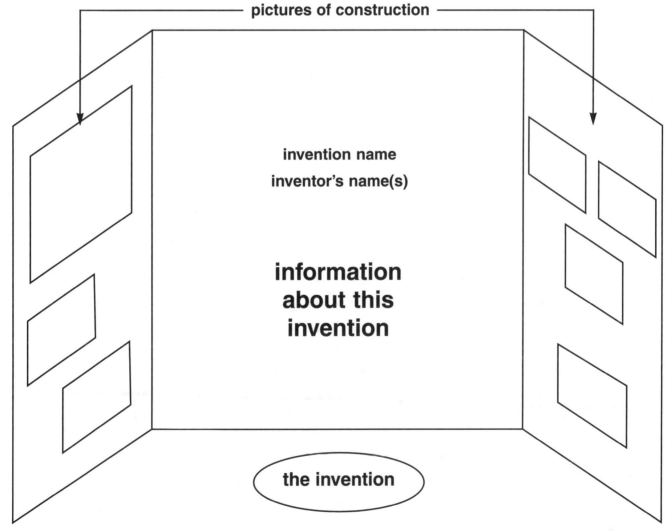

Inventors Convention •

Organizing the Event

If the Inventors Convention is a schoolwide effort, there will be a need for coordination and long-range planning to make it a success. Begin this planning at least three months before the date of the event. This date should be determined by the teachers, since they will play the most important role. This event may be scheduled for a time which is most convenient for your students and their families because you will want the best turnout possible. This may mean that the event is scheduled for after school hours or even a Saturday.

One or two people should serve as coordinator(s) who will be responsible for establishing and overseeing committees and the convention. A teacher and/or administrator from the school should serve as coordinator(s). The coordinator(s) should be free from all other responsibilities during the Inventors Convention since they will be needed to make sure everything is running smoothly throughout the event.

Committees should be formed to spread the work load among many people. The number of members on each committee will vary, as well as the amount of time needed to complete the tasks. Committees should include administrator(s), teachers, parents, and possibly upper-grade students from the school. The committees and some of their responsibilities are described below.

The following information on committees will assist you in organizing this special day. These should be modified to fit your needs.

Program Committee

- Determine the number and types of inventions to be displayed. *(1 month)*
- Find the appropriate location for the exhibits to be displayed. *(1 month)*
- Assign appropriate locations for the exhibits. *(3 weeks)*
- Assign adults to supervise the display areas before, during, and after the event. *(3 weeks)*
- Have each teacher complete a List of Inventions for Convention form (page 68). *(3 weeks)*
- Work with the teachers on assigning the locations for the exhibits. *(3 weeks)*
- Make a map of the display areas, showing where each invention will be located. *(3 weeks)*
- Make name tags for committee members and other personnel involved in the event. Request that teachers make name tags for their students. *(3 weeks)*
- Be sure all teachers are instructing students to make display boards. *(3 weeks)*

Public Relations Committee

- Invite school board members, district office administrators, and other principals. *(1 month)*
- Prepare and send press releases about the event to the media. *(1 month)*
- Disseminate information about the event to families. *(3 weeks)*
- Arrange to have volunteers videotape and photograph the event. These visual records of the event may be used for a presentation to the PTA, school board, or other schools which may want to conduct their own Inventors Convention. *(3 weeks)*
- Contact the media again to issue a reminder of the event. *(1 week)*
- Make videotapes and photos of the event available to the coordinators.

Inventors Convention

Organizing the Event (cont.)

Materials and Equipment Committee

- Create a list of equipment and materials needed during the convention. This may include extension cords and lights. *(1 month)*
- Gather equipment and purchase materials as needed for the event. *(1 month)*
- Assign a person who will be available throughout the day to locate materials and equipment requested by exhibitors. *(1 month)*
- Establish a system for inventors to clean their areas at the end of the event and return borrowed equipment and unused materials to a central location. *(1 month)*
- Make a map of the display area(s) showing where equipment and materials are needed. *(2 weeks)*
- Distribute materials and equipment to the display areas. *(1 day)*
- Deliver materials and equipment to outside display areas. *(2 hours)*
- Check the returned items with the loan list to be sure all have been collected, and then redistribute items to their original locations. *(immediately after)*

Facilities and School Grounds Committee

- Arrange for student supervision before and during the event (possibly teachers). *(1 month)*
- Arrange for adult supervision for students whose parents will pick them up afterwards. *(1 month)*
- Notify school personnel if the event is going to change any regular schedules. *(1 month)*
- Make any necessary preparations for extended parking and traffic control. *(1 month)*
- Be sure parents are notified of the time their children will need to be present for the event and when they will be leaving to return home. *(2 weeks)*
- Organize a system for traffic flow through the exhibit area to avoid "jams." This may require supervising adults to ask people to move forward so all visitors can see the inventions. *(2 weeks)*
- Check all display sites to be sure they are ready for use by the inventors. *(1 week)*
- Check all display areas to be sure the sites have been left in good condition.

Hospitality Committee

- Arrange for greeters and tour guides to assist guests during the event. *(1 month)*
- If appropriate, arrange for refreshments to be served during the convention. These may be complimentary or sold as a fund-raiser to offset the cost of the convention. *(1 month)*
- Make arrangements for a table to be set up at the school entrance where greeters will be stationed with information for visitors, such as a map of display areas. *(2 weeks)*
- Send notes of appreciation to those who helped make the event successful.

Evaluation Committee

- Design an evaluation form for teachers and their students to complete. *(1 month)*
- Distribute evaluation forms to teachers. *(1 week)*
- Arrange for a collection system of the evaluation forms at the end of the event. *(1 week)*
- Make a summary report of the evaluations for the coordinators. Suggestions on the evaluations will be useful when planning to offer this event again.

Inventors Convention •

List of Inventions for Convention

Name of Teacher: _____ Room: _____ Grade(s): _____	
Student Name	**Name of Invention**

Inventors Convention Summary

Since this is a schoolwide event, teachers should assist in some of the responsibilities. They can help with such duties as making name tags for their students. They should also be involved in supervising the students as they construct, set up, and remove their invention displays. Teachers should make a list of their students and the names of their inventions to submit to the Program Committee at least three weeks before the convention. The Program Committee will work with the teachers to assign the locations of the exhibits.

The locations for exhibits may be in a large room such as the cafeteria or in individual classrooms. The outside area may also be used, depending upon weather and time of day. Remember to permit enough space between the exhibits for the inventors to be able to stand near their displays to demonstrate their inventions and to answer questions about them. When setting up the exhibits think of a traffic flow system to allow visitors to get close to them, to see how they work, and then to move on to the next display. The number and size of the exhibits will determine which areas will be most appropriate for the displays. Arrange a grouping of inventions if a large area is used for the convention. This may be by grade level, classroom, or invention types.

The teachers on the committees will be responsible to serve as liaisons with the rest of the staff. Announcements regarding committee activities and teacher responsibilities may be made at staff meetings or by notices sent to the teachers. It is best to give written instructions and information so there are no misunderstandings. Remember to keep the office staff informed of any plans which may require their assistance. Send them a copy of information being distributed to teachers and parents, as well as other details of the event, so they can answer questions parents or others may ask. Be sure this includes a copy of the display map so they can have it for reference.

The Inventors Convention will be a huge success if all members of the staff and committees work together. Perhaps the most important things to remember are to begin the planning early and to develop a spirit of teamwork. Everyone will benefit from this experience if all have, a role in organizing and conducting it, including the students.

Family Source •

Family Science
Overview

When students go home to tell their parents how much fun they are having in their science classes, some parents may question whether they are really learning science. Many parents were taught science through the old system of memorizing facts with little thought given to developing a depth of understanding in regards to scientific concepts. Family Science is designed to involve parents in discovering the excitement and importance of learning science through hands-on activities. It provides an opportunity for parents and extended families to experience how science is being taught and to develop support for the technique of teaching science through direct involvement. Parents may be surprised to see how much science can be learned through hands-on activities as opposed to just memorizing facts. Their children will be working with them during these Family Science events, proving to the parents how much progress the students have made in the subject.

There are a variety of ways to conduct Family Science events which provide opportunities for students to share their knowledge and enthusiasm for science. Here are a few of them.

- **Family Science (pages 71–77):** Children and parents become scientifically literate by doing science investigations together. This may be done in individual classrooms or throughout an entire school. Information on how to conduct such an event and sample activities are provided.

- **Family Nature Walks (pages 78–81):** Using wildlife areas near the school or within an easy driving distance, families have the opportunity to learn science as they experience a carefully planned walk. Descriptions for conducting eight different walks are included. They are bird, farm, insect, microworld, moonlight, night, wetlands, and wildlife sanctuary walks.

- **Evening Science (pages 82–85):** Arrangements are made for families to attend programs at local science facilities which may include a museum, planetarium, observatory, zoo, or aquarium. Group discounts help give parents a price break. Families will be introduced to the advantages of visiting each facility and will receive additional knowledge and appreciation for science.

- **Science Performances (pages 86–93):** Another method for students to learn science is through creative dramatics. This can be done through a variety of science performances presented for their families. Ideas included in this section are the following:

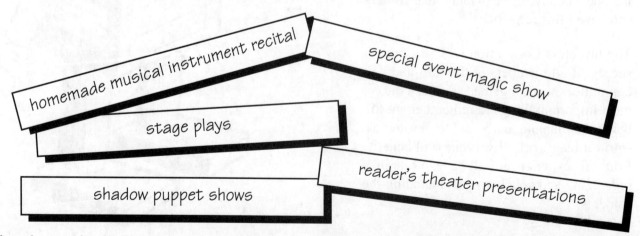

- homemade musical instrument recital
- special event magic show
- stage plays
- shadow puppet shows
- reader's theater presentations

When the curtain closes and the science equipment is packed away, students, families, and the school staff will all benefit from the experiences they have had at Family Science events.

Family Science (cont.)

Overview (cont.)

What Is It?

For several years, the University of North Carolina at Chapel Hill has served as a training site for EQUALS Family Science Programs. (See Resources, page 96.) EQUALS is a national outreach program designed to teach science by having K–8 students and parents learn and enjoy science together. The goal of Family Science is for children and parents to become more scientifically literate by doing science investigations together.

Who Is the Audience?

The program is especially designed for parents who might not typically attend parent-teacher conferences or school functions. However, all parents and students are encouraged to attend and benefit by experiencing exciting science together. During the Family Science events, "family" consists of a child and a caring adult. The adult may be a parent, neighbor, older sibling, grandparent, or community "big brother/sister." Thus, students whose parents may not be able to accompany them to the event can attend with another adult. Having an adult accompany the child helps with supervision and provides a companion and audience for the child. The goal is for the investigation or activity to remain child centered. The adult is encouraged to let the child take the lead during the activity session.

Where and When Should It Be Held?

Family Science programs may be conducted after school hours at the school, community center, local museum, church, or other area which is convenient for the program and audience. Some parents may be unable to come to the school; thus, some of the alternatives listed earlier may be a good location for some families. Using the school as the site for the Family Science event will make it easier for the coordinators if they are school personnel. Classroom and cafeteria space will be available, and there will be no charge for this activity. The activities may be set up in various classrooms or centrally in the cafeteria. A rotation system can be designed to move the groups through several activities during the event.

If the program is conducted after school, refreshments may be a good idea since some parents may have come directly from work without having had time to eat. Also, eating together frequently builds bonding among strangers and thus will make a better environment for the activities.

A variety of days and times may be used for the Family Science event to enable more people to attend, if it will be offered more than once during the year. This may include evenings, Saturdays, the extra day of long holiday weekends, or during winter or spring break.

Family Science •

Family Science (cont.)

Overview (cont.)

How Should the Event Be Conducted?

Adults and children will be working together in small groups. This will mean they need to be ready to share responsibilities while doing the activities and to discuss their ideas and discoveries as they progress. It is, therefore, important to develop a cooperative learning atmosphere for the event. This may be accomplished by the way the activities are structured and how they are located in the facility being used. Spaces between activity centers will be needed to enable each group to work without interfering with the progress of another group.

If the school has a large population, it may be wise to invite only part of the school at one time to the event. The division may be done by grade levels K–2, 3–4, and 5–6 in order to restrict the number attending and to limit the number of activities needed.

How Should Activities Be Structured?

The key to success for Family Science is to keep in mind the variety of backgrounds the participants will have. Many of them may not feel comfortable with the science field, while others may be professional scientists. It is important to know the level of the audience and structure the activities to match their skills. To play it safe, provide activities which do not require technical or scientific backgrounds; in other words, give enough information for a non-scientist to participate.

Emphasis should be placed on exploration and self-discovery, much in the way classroom activities are designed for students. The activities should not require a right answer but rather the opportunity to learn by doing, and they should encourage various techniques of exploration.

Remember that adults will have various levels of reading and comprehension skills. Keep the written information, such as background materials and instructions, simple and to the point. Have adults serve as monitors to assist as needed without taking charge of the activity.

Each activity should teach only one science principle or "big idea" to avoid overwhelming the participants. Above all, the activities should leave adults and children with feelings of success and enthusiasm about science.

How Can a Family Science Event Be Organized?

Other sections of this book have provided details for conducting a Science Discovery Day and an Inventors Convention. Many of the coordination details for those apply to Family Science, as well. Coordinators and committees should be established to make Family Science a success without overburdening one or two people. The community can also offer support from many different groups within and outside the school and offer a wealth of talents and ideas.

Groups may rotate through more than one activity by the use of tickets or by establishment of a time limit for the centers and designation of how and where the groups should move. Remember to give each group time to clean up its center before moving on to the next one. The appropriate rotation system will depend upon the structure of the activities.

•• *Family Science*

Activity Ideas

What Science Activities Should Be Included?

There are various resources which can be used to find appropriate activities for a Family Science event, and some of these are listed in the Resources section of this book (pages 94–96). Many of the activities outlined in the Science Discovery Day and Inventors Convention chapters of this book may be used for Family Science as well. Additional activities are included below and on the next two pages.

What Is It?

Goal: Teams of children and adults will investigate unusual objects.

Materials: Unique and unusual items such as the following:

- bones from an owl pellet
- fruit (e.g., pomegranate)
- seedpods
- tools
- toys
- index cards or notepaper
- small paper bags

Preparation:

- Place each item with tools needed to examine it, such as a magnifier, in a paper bag.
- Include a simply written set of instructions which will be used to initiate the investigation but not give answers. These may include hints on how to proceed with the investigation and questions which will lead to discoveries.
- Number the bags and keep a list of the contents of each bag.
- Place the bags on a large table.

Procedure:

- Divide the participants into small groups of no more than six people.
- Let the groups select one item each to investigate. As they finish with their items, they should return them to the tables for another group to use and take a new bag. Provide each group with lined index cards or notepaper to record their findings so they may be shared at the end of the event.

Closure:

At the end of the program, have the groups compare information about the items investigated.

Family Science

Activity Ideas (cont.)

Mystery Scientists

Goal: Determine a science career by observing the tools used on the job.

Preparation:

- Invite at least five people from various science careers in the community to be guest presenters at a Family Science event and to share science tools they may use in their jobs. Try to select people who break the stereotypes of people in science and technical careers such as a male nurse who can show instruments, such as a stethoscope, or a farmer who can show instruments used in dehorning cows or other equipment which would probably be unfamiliar to the audience.
- Request that each guest speaker come dressed in the clothing he or she wears on the job. This will provide another hint to the audience.
- Consider videotaping the presentations for use with the PTA or other groups who might be helping to fund the Family Science programs.

Procedure:

- Divide the participants into small groups and have each group select a recorder and spokesperson.
- Have the guest speakers show their science instruments without giving any hints about what they do as a career. Each recorder should take notes of clues that are given during the presentation.
- After each presentation, provide time for the groups to discuss what they saw and heard and to tell what they think the speaker does as a career.

Closure:

- Have each presenter tell what he or she does for a career.
- Let the audience ask questions regarding their work and especially the uses of the tools shared.
- Serve refreshments so the participants can view the science instruments again and informally talk with the guest speakers about their careers.

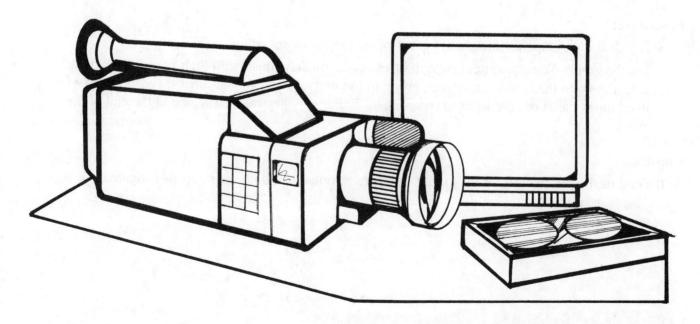

#2509 Beyond the Science Fair　　　　　74　　　　　© Teacher Created Materials, Inc.

Family Science

Activity Ideas (cont.)

Is This a Seed?

Goal: Learn to distinguish seeds from non-seeds.

Materials:

- variety of seeds which include peanuts, seeds from fruits and vegetables, large lima beans, spices, and wild seeds
- 25–30 packages of seeds with pictures of the plant on them
- items which resemble seeds such as buckshot
- index cards
- white glue
- self-sealing plastic bags
- large paper plates
- magnifiers
- soaked lima bean seeds

Preparation:

- Fill bags with a variety of seeds and non-seed items so they all have samples of the same items.
- Glue several seeds from each seed package to individual index cards. Number the cards. Make a record of each seed card number and the name of the plant. Make a list of the plants on the packages and copy the list to be used as an answer sheet for the closure activity.
- Glue several seeds from each package onto the outside of the package near the picture of the plant. Seal the package so it cannot be opened.
- Divide the paper plates into thirds with a felt pen. Label one section "yes," another one "no," and the last one with a question mark.
- Soak enough lima bean seeds for every participant to get one. This is done by placing the seeds into hot water (not boiling) for about an hour so they will soften and expand.

Procedure:

- Divide the participants into small groups and distribute a bag of seeds, a magnifier, and a paper plate to each group.
- Explain that they need to sort the material in the bag into one of three categories on the paper plate. The categories are "yes" for those which they know are seeds, "no" for those which are not seeds, and "?" for those about which they are uncertain.
- When they are finished, have them discuss what criteria they used to sort the seeds.
- Let the groups compare how they sorted the seeds.
- Ask them how they can be sure they are seeds (e.g., plant them to see if they grow).
- Show them another way to tell seeds from non-seeds by distributing the soaked lima bean seeds. Have them peel off the outer skin and gently open the seeds. They will see two tiny white leaves and the beginning of a root inside near one edge of the seed. Have them examine this with their magnifiers to see the veins on the leaves.
- Instruct them to open the peanuts as well, and they will see the same embryo plant inside. Explain that all seeds have this inside of them. This is another way to prove something is a seed.

Closure:

Distribute the index cards with the seeds glued on them.

Family Science •

Center Activities

The following activities are to be set up in centers where participants can work and then rotate after a given time to a new center. The materials and instructions should be set up at each center. Each center will accommodate more than one adult-child team. Duplicates of the centers may be made in order to spread them into a larger area. Depending upon the time available and the complexity of the centers, participants may rotate to three or four centers during the event.

Funny Pictures

Goal: Discover the properties of reflection in mirrors.

Materials per Center:
- mirrors without edges covered
- colored photos from magazines
- kaleidoscopes (optional)

Procedure:
- Lay a picture on the table.
- Hold one mirror perpendicular to the picture.
- Slowly pull the mirror down the picture, watching the reflection as it creates double images.
- Experiment with turning the mirror over the picture to see how the image changes.
- Place two mirrors together so the edges touch, creating a V shape.
- Repeat the third step to see how two mirrors change the image.
- Place the two mirrors on top of one section of the picture and slowly open and close the mirrors like a book. Watch the image to see what happens. Do you see more images?

Try this on several other photographs. You have just discovered how a kaleidoscope works.

Colorful Drops

Goal: See what colors result when primary colors are mixed.

Materials per Center:
- dropper bottles of red, green, blue, and yellow food coloring
- sheets of waxed paper
- cotton swabs
- white paper
- papers with six circles on each

Preparation:
- On sheets of white paper the same size as the sheets of waxed paper, draw six circles the size of a quarter. Space these in such a way that they will form two rows of three circles each. Write the color combinations below them, for example, red + green.

Procedure:
- Lay a paper with circles on it on the table and cover it with one sheet of waxed paper.
- Place a drop of color in the first circle which matches the color name. Add a drop of the second color beside the first drop. Use a cotton swab to mix these two colors.
- Use the cotton swab to paint the new color onto a piece of white paper. Write the name of the new color beside the paint.
- Repeat steps two and three with all three colors. What new colors did you make?

Family Science

Center Activities (cont.)

Mystery Smells

Goal: Use the sense of smell to identify various aromas.

Materials per center:
- 10–15 opaque, empty, film canisters
- masking tape
- cotton balls
- awl
- hammer
- answer sheet inside an envelope
- guess sheet (several copies)
- various liquids with strong aromas, such as vinegar, vanilla, peppermint, or coffee
- permanent pen

Preparation:
- Use the awl to put a large hole in the lid of each film canister. Cover the hole with tape.
- Put a piece of tape on the side of each canister. Use a permanent pen to number each canister on both tapes so the lids and canister numbers are the same.
- Into each canister drop an individual cotton ball that is saturated with an aroma. Cap the canister. Each container should have a different scent.
- Make an answer sheet by listing the numbers and writing the names of the scents beside them. Place a copy inside an envelope marked "Answers."
- Make a Guess Sheet by listing the numbers of the containers with a line beside each number.

Procedure:
- Select any canister and circle its number on the Guess Sheet. You do not need to go in order.
- Carefully lift the tape and smell what is inside. (Do not open the lid!)
- Write what you think it smells like on the line beside the number.
- Repeat this for all containers. When finished, take the Answer Sheet from the envelope and check to see how many aromas you guessed correctly.

What Do You Feel?

Goal: Use the sense of touch to identify objects.

Materials per center:
- 5–6 small coffee cans
- large socks with elastic ribbing
- guess sheet
- pencil
- variety of items to place in cans (ball, scissors, paper clip, etc.)
- wide packing tape
- answer sheet inside an envelope
- permanent marker

Preparation:
- Cut the cuffs from the socks and place them over the coffee cans. Secure them in place with the tape.
- Place a different item in each can. Place a challenging item such as an apple in one can.
- Number the cans with a permanent marker.
- Make a list of the contents on an answer sheet. Place this in an envelope marked "Answers."
- Make a Guess Sheet by numbering it and leaving room for a guess to be written and drawn.

Procedure:
- Select a can and circle its number on the Guess Sheet. You do not need to go in sequence.
- Carefully put your hand inside and feel the item in the can. (Do not peek inside!)
- Write what you think it feels like on the line beside the number. Make a drawing of the object.
- Repeat this for all cans. When all are finished, take the Answer Sheet from the envelope and check to see how many items you guessed correctly.

© Teacher Created Materials, Inc.   #2509 Beyond the Science Fair

Family Science •

Family Nature Walks

What Is It?

This is an opportunity for families to visit the wildlife areas of your community on organized nature walks. It is a chance for students and their families to extend their science learning in an outdoor laboratory setting. Here, they can observe and investigate flora and fauna which they may have otherwise overlooked. Once families have been involved in such enjoyable encounters with the nature in their neighborhoods, they may be inspired to go on nature walks of their own.

Who Is the Audience?

The nature walks should be open to all ages of children and adults. The children should be accompanied by an adult to help with the supervision and activities. The ages and physical abilities within the groups may determine how strenuous the walks are. Some walks in rough terrain may limit the groups to those who are able to hike several miles. Consider doing more than one nature walk in easy to difficult areas so various ages and ability levels can participate.

The groups can be small or large, as long as there are many leaders available to keep the leader/participant ratio at about one to ten. There may be three or four children to an adult if necessary. Be sure to send home information that specifically tells families what to expect and how to prepare for their walk. (See the sample announcement on page 81).

Where Should It Be Conducted?

The sites for the walks may be in a local forest, pond area, field, or park. More than one site may be used to offer a variety of walks from which families can choose. These may be on the same date or spread over the year. It will depend upon the number of leaders who are available. Seasonal changes will take place so that a site may be visited more than once in the year. It can be developed into a research study for comparing the changes in one site throughout the year.

When Should It Be Conducted?

The walks can take place throughout the year in all kinds of weather and at various times of the day. This will provide a wide range of observation conditions and opportunities to see the animals and plants under different conditions. Some animals come out at night (nocturnal), while others move about during the day (diurnal). Some plants are also diurnal, opening when the sunlight touches them. Others are nocturnal, giving off a scent as it gets dark. Rainy or cold weather will also affect animals and plants in their behavior as well as appearance.

How Should It Be Organized?

Be sure the leaders tour the sites at least twice before the outing. This will enable them to find areas of interest and to develop a route. The success of a walk depends upon careful planning so that every possibility is considered, including troublesome fauna (e.g., snakes or stinging insects).

Family Nature Walks (cont.)

What Activities Should Be Included?

Here is a list of possible walks.

Bird Walk: Early in the day is the best time for this walk since birds are more active then. The best time of year is when birds are migrating. To observe birds in southern latitudes, walk in the late fall when they are migrating south. It is just the opposite in northern latitudes when the best viewing is in the spring as the birds return home. Bring binoculars and bird identification books. Wear dull colors to blend with the native plants and avoid alerting the birds to your presence.

Farm Walk: This is an opportunity to take families to a farm. They can observe the important work that goes into raising cattle, milking cows, or tending chickens, pigs, and other livestock. Take small groups so they can get a personal tour and perhaps even help to milk a cow or feed the livestock.

Insect Walk: This needs to be done during the warm part of the day since most insects tend to move very little when it is cold. Take magnifiers and self-sealing plastic bags. Place live specimens in the bags for examination and then release them. Take crackers to crunch into crumbs and sprinkle along ant trails. Observe the "before and after" of adding the food. Watch to see what they do when their trail is interrupted by drawing a line across it with a stick or placing a large pebble in the path. Bring along insect identification books.

Microworld Walk: Use magnifiers to discover another world of tiny plants and animals. Place a loop of string around an area, such as the base of a tree, and examine everything within the circle, using the lens to find the details. Collect specimens for others to examine. Release these at the same spot after studying them. Nothing should be harmed or removed.

Family Science

Family Nature Walks (cont.)

What Activities Should Be Included? *(cont.)*

Moonlight Walk: A good time for this is during a full moon. The full moon rises as the sun sets, permitting good illumination when the sky gets dark early in the evening. Winter is the best time for a moonlight walk since the hours of daylight are shorter. A few days before and after the full moon also offers enough light to see by if the sky is clear. This is a great opportunity to see animals such as owls and bats and also to hear coyotes or other night hunters in some areas.

Night Walk: This is a chance to observe nocturnal wildlife. Walk without making any sounds and stop frequently to stand quietly and listen for sounds. Animals will slowly begin to move about. This takes some time, so if the group sits down and remains quiet, there is more opportunity to hear and to see the animals. Bring flashlights to use but do so sparingly. Covering the lens with red cellophane will help to dim the light so you can see, but the light will not illuminate the entire area and frighten away the animals.

Wetlands Walk: This can be done in any season. Weather will dictate the type of clothing and time spent exploring the area. It is worthwhile to schedule visiting this area during all four seasons. Even in locations in southern California where seasonal changes may be hard to detect, wetland plants and animals are definitely affected by these slight changes. Wear waterproof boots and layers. Carry a clean jar with a lid and a small fishnet such as those used in aquariums. These can be used to collect water insects, small fish, tadpoles, and other organisms that will be placed the jar temporarily while they are being examined and then released. Bring binoculars to look for wetland birds and identification books for birds, insects, and reptiles.

Wildlife Sanctuary Walk: Docents may be available at these locations to take the families on a tour. Their expertise can help to make the trip more meaningful for all involved. Be sure to take the docent tour first to see if it is appropriate for your families. Ask if maps or other literature will be made available to each family so they can return on their own at another time.

#2509 Beyond the Science Fair © Teacher Created Materials, Inc.

•• Family Science

Sample Nature Walk Announcement

Use this announcement as a template to create your own. If desired, simply white out the areas that do not pertain to you and type in your own information.

Family Nature Walks
Sponsored by Orion School
Information and Reservation Form

What: We are planning to conduct a number of nature walks throughout this school year. These will give our students an opportunity to investigate a variety of natural settings in our greater neighborhood. The goal is to extend science beyond the classroom and into the wild.

Who: The walks will be open to all students and their families but not all our student population can attend at the same time. We need to limit the sizes of the groups to provide a quality program. Selection of attendees will be on a first-come-first-serve basis so be sure to get your reservation form in early. Every child must be accompanied by an adult. This may be a parent, high-school-aged or older brother or sister, grandparent, or adult friend.

Where and When: Our plans are still being made, but we hope to offer the following series of walks this school year. Our schedule is tentative, and your suggestions are welcome. *(List the nature walk locations you plan to visit. Include the date for each walk beside the name of the site. See the example below.)*

Wetland Walk at Jenny Pond   Saturday, October 16, 1999   8:30 A.M.–2:30 P.M.

We will carpool to Jenny Pond from the school, leaving the parking lot at 8:30 sharp. While at the pond, we will use small fish nets and jars to collect water insects and other small water life. These will be returned to the pond after we observe them. We will walk the paths around the pond area to look closely at plants and animals in that area.

For more information, call the school at (555) 555-5555.
Important: Wear waterproof boots and bring a sack lunch with a beverage.

- -

Reservation Form
Wetlands Walk: Jenny Pond
Deadline: Wednesday, October 13, 1999
limited to _____ participants

If you are interested in joining us on this nature walk, please print the following information and return this part of the form to school by the deadline. You will receive a confirmation.

Child: _____ Adult: _____ Relationship: _____

Home Phone: (   ) _____

If you can offer transportation for others, check the following box.

☐ I am willing to drive in a carpool and have _____ extra seats (with seatbelts) for passengers.

Signature: _____

© Teacher Created Materials, Inc.   #2509 Beyond the Science Fair

Family Science •

Evening Science

What Is It?

This is an opportunity to involve families in a visit to a local science museum or other facility. Group rates are available at most of these. Such excursions will allow not only the science experience to be extended beyond the school, but also parents will learn the value of visiting these facilities and may return on their own to attend future programs.

Who Is the Audience?

All students and their families may attend these programs. Since admission is charged at most of them, some families may find this a financial hardship. Check with the PTA to find out if it is possible to offer full or partial scholarships to families unable to pay the admission. Partial scholarships would require the family to pay a percentage, thereby spreading the fund further. Parents may also be willing to contribute to a fund by adding to their admission price when they make their reservations. Be sure to send thank-you notes to those who contribute.

Where Should We Go?

Contact the various science facilities in your area to get their schedule of events and group rates. The facilities you visit will depend on your location. Some suggestions are listed below.

- planetarium
- natural history museum
- aquarium
- zoo
- science and industry museum
- aerospace museum

Check with less obvious science sites such as an observatory, college planetarium, hospital, or research center to see if they offer public tours. Select programs that coincide with your school's curriculum. If a program is not suitable for all age levels, offer it only to the appropriate grades.

When Should These Be Offered?

Schedule the Evening Science for weekends to avoid late nights with school the next day. You may even consider offering an Evening Science event during a vacation period.

How Should This Be Organized?

Needed for this event are coordinators who will be responsible for the following tasks:

- Locating appropriate science programs and scheduling them for the school year
- Designing and distributing announcements
- Collecting money and reservations forms from families
- Making reservations at the science facilities
- Sending confirmations and information to those who will attend
- Arranging scholarship funds for families who need financial assistance
- Arranging carpools or other transportation (optional)
- Accompanying the families to the event to pay the bill and assist as needed

Examples of an announcement, a confirmation, and a waiting list notice are provided on pages 83–85.

Family Science

Sample Evening Science Announcement

Use this announcement as a template to create your own. If desired, simply white out the areas that do not pertain to you and type in your own information.

Evening Science
Sponsored by Orion School
Information and Reservation Form

What: We are planning to offer a series of Evening Science programs during this school year. This will give our families an opportunity to discover the exciting science programs being presented at these local facilities.
(List the facilities you plan to visit this year. If available, include the date for each event beside the facility name.)

Who: The science programs will be open to all families. Some programs may be appropriate for older or younger children. If this occurs, attendance will be restricted to specific grade levels. Some facilities may have space limitations. In this case, reservations will be on a first-come-first-served basis until the quota is filled.

Limited funds have been established by the PTA for partial scholarships for those who may need this assistance. In order to stretch the scholarship funds, we will offer fifty-percent reductions to those who qualify. If you wish to apply for a scholarship, please check the box on the reservation form. Contributions to the scholarship fund are gratefully accepted. Please include the amount in your reservation check and enter the amount of your contribution in the total.

Where: Our first Evening Science event is shown on the reservation form. Please complete this form and return it to school by the deadline. Space is not limited for this event, but we must confirm our reservations three days before the event.

For more information, please call the school at (555) 555-5555.

Important: Children must be accompanied by an adult.

Evening Science Reservation Form
(name of the event site)
Show: (title of the program)
(day and date of the event)
Program Begins: (beginning and ending time)
Deadline: (three days prior to the event)

Print the following information and return this part of the form to school by the deadline.

 # Tickets   Price   Total

_____Child x (cost per ticket) = _____   (If you are applying for a scholarship, fill in this
                                              information but do not send a check.)

_____Adult x (cost per ticket)   = _____
       Total Amount   = _____

Make checks payable to: (name for check)
Home Phone: (   )
Child's name: _____ Adult's signature: _____

❏ I would like to be considered for a partial scholarship. (You will be contacted.)
❏ I am including $_____ as a contribution toward the Evening Science Scholarship Fund.
❏ I am willing to drive in a carpool and have _____ extra seats (with seatbelts) for passengers.

Family Science •••

Sample Evening Science Confirmation

Use this form as a template to create your own. If desired, simply white out the areas that do not pertain to you and type in your own information.

Evening Science Confirmation
(name of event site)

Show: (title of program)
(day and date of the event)

Program Begins: (beginning and ending times)

Name: _____

Reservations: _____ children and _____ adults

We are pleased to confirm your reservations for the Evening Science event. Your tickets will be distributed when you arrive at the site. Please remember the following:

- Bring this confirmation with you.
- Arrive at least thirty (30) minutes before the program begins.
- Cancellations cannot be accepted after *(day, date, and time)*.

Directions to *(name of science facility)*:

(Write the directions and if possible, include a map to the location, including parking areas.)

If you will be driving in a carpool, meet in the school parking lot by _____. *(Designate a time which allows families to arrive at the school parking area, team up in cars, and drive to the facility.)* All cars will leave the parking lot by _____ sharp. *(Designate a time that permits ample driving time to the facility.)*

Questions? *Call the Evening Science coordinator at (555) 555-5555.*

#2509 Beyond the Science Fair © Teacher Created Materials, Inc.

Sample Waiting List Notice and Scholarship Fund Acknowledgment

Use these forms as templates to create your own. If desired, simply white out the areas that do not pertain to you and type in your own information.

Evening Science Waiting List Notice
(name of event site)
Show: (title of program)
(day and date of the event)

Name: _____

We regret to inform you that, due to limited space, not all reservations could be accepted. Tickets were assigned on a first-come-first-served basis until all were distributed. Your reservations were among those which could not be confirmed at this time.

Your name and ticket request have been placed on a waiting list. At this time you are #_____ on the list. If there are any cancellations, we will select replacements from this list. You will be notified as soon as possible if your reservations are accepted. If we are not able to provide tickets for you, your check will be returned to the address which appears on the check.

We hope this will not discourage you from making reservations for future Evening Science events.

If you have any questions, please call the Evening Science coordinator at (555) 555-5555.

- -

Evening Science
Scholarship Fund

Acknowledgement and Thanks

Thank you very much for your contribution of $_____ to the (school name) Evening Science Scholarship Fund. It is the generosity of contributors like you that makes it possible for us to extend financial assistance to our families who need it in order to attend these special events.

We hope you will enjoy the Evening Science events, especially with the knowledge that you helped make it possible for someone else to attend as well.

Most sincerely,

_____
(coordinator of scholarship fund)

_____
(school principal)

Family Science •

Science Performances

What Is It?

A Science Performance provides the opportunity for individual classes or groups of students to put their study of science into an art form.

Who Is the Audience?

Performances may be given for another class or as a special assembly for the school or PTA. Parents should be invited to attend, to give them the chance to observe science in a different form. They will also develop an appreciation for their children's skills in science and language arts.

What Should Be Presented?

The Science Performances may be in the form of a play, recital, or reader's theater. They may be based on a topic being studied in science, related to a special event or holiday, or portray the lives of famous scientists. These may include the following:

- constructing homemade musical instruments and presenting a recital
- doing a science magic show for a special event such as a birthday
- presenting a space play about a trip to the moon
- offering a reader's theater about the lives of famous scientists
- performing a shadow puppet show about a science topic

Ideas which can be used for these various types of performances can be found on pages 87–93.

How Will It Be Organized?

The organization needed for the performances will depend upon how extensive the audience is going to be. A presentation to another class or just to the classroom parents may be sufficient and will take very little organization. If the performance is a play, it may require stage settings, costumes, lighting effects, and music. This will require more time to organize, including reserving the auditorium where it will be presented and scheduling practice sessions for the cast.

Students may be enlisted to help create announcements about their performance(s). This may be done in the form of flyers, posters, letters of invitation, and other ways to spread the word.

If this will be a play, check for events which may already be on the school calendar before scheduling the performance. This will help to avoid competing with others for space and an audience. Place the date and time for the performance on the school calendar as soon as possible to ensure the use of the location and equipment required for the play.

Request the assistance of parent volunteers if costumes are needed or to assist with rehearsal sessions. It is always good to have the aid of additional adults during this busy time.

• Family Science

Science Performance Ideas

Homemade Musical Instrument Recital

Overview: Musical instruments are constructed from a variety of materials.

Materials:

- glass bottles
- oatmeal boxes
- large balloons
- waxed paper
- 12 drinking straws
- water
- shoeboxes
- packing tape
- combs
- long steel nails
- rubber bands of various widths
- 12" (30 cm) long, thin wood strips
- large and small coffee cans
- lyrics to simple songs such as "Row, Row, Row Your Boat" (optional)

Procedure:

Create musical instruments as follows:

- Fill the bottles with water to different levels. Tap on these with a steel nail and arrange them in order of pitch.
- Sort the rubber bands by their widths. Place them in order from narrow to wide around the long side of a shoebox. Pluck the rubber bands to hear their tones and arrange them, as necessary, to have them play a scale.

- Cut six to eight wood strips to various lengths, making each one about one inch (2.5 cm) shorter than the last. Tape these in order of their lengths to the edge of a table. Pluck them in order to hear the pitch. Arrange or trim them as necessary to play a scale.
- Cut the balloons open and stretch them over the open ends of the oatmeal boxes and cans. Secure them with packing tape. Tap on them as drums.

- Fold a piece of waxed paper over the teeth of a comb. Place this fold and the comb lightly between the lips and blow. This will produce a sound. (It may also tickle the lips!)
- Cut the drinking straws to different lengths and sort them by length. Lay them flat and secure them with clear tape. Blow across them to be sure they are in order of pitch. Rearrange them as needed to make a scale. (*Variation:* Do not tape the straws together but let them be played by individual students.)

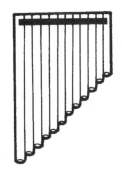

- Let students play tunes on the instruments individually. Have them combine some or all of the instruments to play the same tune.

Have the students invent additional instruments and include them in the performance.

© Teacher Created Materials, Inc.

Family Science

Science Performance Ideas (cont.)

Special Event Magic Show

The following science "tricks" may be performed as a magic show. See the Resources section (pages 94–95) for suggestions of books which have additional activity ideas.

Red, White, and Blue

Materials:

- aluminum foil
- ¼ cup (65 mL) milk
- ¼ cup (65 mL) clear corn syrup
- red food coloring
- ¼ cup (65 mL) lamp oil (caution: flammable liquid)
- clear, narrow glass or plastic cup
- blue oil paint (available in artist supply stores)

Preparation:

- Color the lamp oil dark blue by mixing in the oil paint.
- Color the corn syrup dark red by mixing in the red food coloring.
- Cover the containers with aluminum foil so that it can easily be pulled off.

Procedure:

- Place all containers on the table so the audience can see them. Tell them you are going to pour all three liquids into the glass. Ask them to predict what color will result when these are combined.
- Slowly pour the three liquids into the glass to avoid much mixing. Pour the red syrup, then the milk, and finally the lamp oil into the glass.
- Remove the foil. The audience will see they have not mixed but formed three colored layers.

Closure:

- Stir the liquids and watch to see what happens. (They will separate into the same three layers.)
- Explain that these liquids are of different viscosities and densities and therefore do not mix.

Soda Pop Float

Materials:

- aquarium
- balance beam scale and gram masses
- 16 oz. (450 g) cans of diet and regular soda pop

Preparation:

- Fill the aquarium with water until it is deep enough to float a can of diet soda pop upright.
- Put the soda pop cans into the water. Arrange them so that the diet ones are bottoms up and the regular ones bottoms down. Do this before the audience arrives so they do not see the pattern.

Procedure:

- Ask the audience to look for patterns (e.g., which cans float, which do not).
- Explain that all cans have the same amount of liquid. Ask for explanations as to why some float and others sink.

Closure:

- Explain that the regular soda contains sugar which adds density to the liquid; thus, it sinks. The diet soda has an artificial sweetener which has much less density; therefore, it floats. The amounts of liquid and gas in each can varies from batch to batch.

Family Science

Science Performance Ideas (cont.)

What Color Is This?

Materials:

- red cabbage
- uncoated laxative tablets
- fork and plate
- ammonia
- test tube holder (can be created with holes punched in the bottoms of foam cups)
- 4 test tubes, olive jars, or 1-oz. (25 g) clear plastic cups
- alcohol
- vinegar
- 3 clear droppers
- water

Preparation:

- Chop about one-fourth cup (65 mL) red cabbage and put it into a microwave-safe container. Cover the cabbage with two cups (500 mL) water and boil for ten minutes in the microwave. Pour off the water.
- Crush five laxative tablets with a fork on the plate. Scrape the crushed material into a clear cup. Pour in one cup (250 mL) alcohol and stir until most of the material is dissolved. Let it set for five minutes until the undissolved material drops to the bottom. Drain off the clear liquid.
- Pour the cabbage juice into three test tubes, filling them three-quarters full. Pour the alcohol mixture into two test tubes, filling them three-quarters full.
- Place ammonia, water, and vinegar in different droppers. Mark them so you will be able to distinguish one from the other.

Procedure:

- Show the audience the test tubes and point out the color of the liquids in them. Show them the three droppers so they may see that they are all clear liquids. Ask them to watch what happens when you place drops of the clear liquids into the test tubes.
- Place several drops of vinegar into the first tube of cabbage juice. Shake it gently until there is a color change. (It turns violet.) Place several drops of ammonia into the next tube and shake it. (It becomes green.) Place several drops of water into the third cabbage juice container. (It remains blue.)
- Place several drops of vinegar into the alcohol mix (no change). Repeat this with ammonia (bursts of bright pink appear) and then add several drops of water. (There is no change.)

Closure:

- Ask the audience what they think is happening. Explain that you are using indicator dyes to determine if something is an acid, base, or neutral. Show them the droppers and explain that one is vinegar (acid), another is ammonia (base), and the last is water (neutral). The cabbage dye turns violet when an acid is added, green with a base, and has no reaction to a neutral substance. The cabbage juice can indicate an acid or base substance by its color change.
- Demonstrate this again by pouring out the cabbage juice, rinsing the test tubes, and then adding new cabbage juice. Tell them what you are adding as you place drops of acid, base, and neutral liquids into the different test tubes.
- Explain that the clear liquid in the test tube is an alcohol mix containing phenolphthalein from the laxative. It is a base indicator so it turns bright pink with ammonia. Add drops of vinegar to show no change, then ammonia to show bright pink, and finally water to show a neutral reaction.

Family Science

Science Performance Ideas (cont.)

Anti-Gravity

Materials:

- powerful magnet (from audio speaker or electric motor)
- table
- tablecloth which will drape to the floor
- large box or smaller table
- 3' (90 cm) clear-looking thread
- large paper clip
- thick book
- piece of glass from a picture frame
- piece of wood
- aluminum pie pan

Preparation:

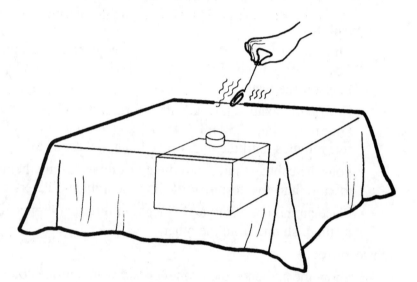

- Use the box or small table to hold the magnet beneath the tabletop so that it touches the table.
- Drape the tablecloth over the table so that it goes to the floor and hides the box.
- Tie the paper clip to the thread. Hold it above the magnet to see if it will be attracted to the magnet without touching the table top. If not, adjust the magnet until it can do so. The paper clip should appear to be suspended in midair.
- Practice the trick of placing your hand between the paper clip and table top. The paper clip should remain at an angle. Do the same with the book, glass, plastic, wood, and aluminum. All of these should permit the paper clip to hang at an angle. The thickness of the book or wood may need to be changed if they put too much space between the magnet and paper clip.

Procedure:

- Show the audience the paper clip on the thread. Hang it over the table beyond the magnetic field of the magnet so that it hangs straight down.
- Say some magic words as you gradually shift the clip position to be over the magnet. Slightly tug the string sideways so the clip hangs at an angle. The paper clip is defying the laws of gravity!
- Tell the audience that you will try to restore the gravity force. Ask a member of the audience to assist you. Have him or her slip a hand beneath the paper clip. (It should remain at an angle.)
- Have the volunteer place the book and then the other items, one at a time, beneath the paper clip.

Closure:

- Say another magic phrase and pull the paper clip higher, beyond the magnetic field, so that it hangs straight down.
- Ask the audience to explain what happened. Draw aside the tablecloth to show the magnet underneath the table

Family Science

Science Plays

What Are They?

Science plays may be presented in many forms. Students may follow scripts which they have written or some which are available in books. These may take the form of a stage play with appropriate props, lighting, and costumes. They may also be a shadow puppet show, simulated television program, or reader's theater.

Who Is the Audience?

The audience should be families of the performers or all the families in the school. This will depend upon the size of the school and the individuals' abilities to appear before a large audience.

Where Should They Be Conducted?

Use a stage, if one is available, to enhance the performance by adding an air of professionalism and importance to the play. Lighting may also be needed, depending upon the performance.

What Plays Should Be Presented?

Examples of a variety of performance ideas are briefly described below. Most of these appear in other books written by the author. See Resources (pages 94–95) for further information.

Stage Play: *You Are Go for Launch* (from *Space: Intermediate* by Ruth Young)

This is a simulated space trip launching from Kennedy Space Center, Florida, on July 20, 2021, 50 years after the landing on the moon. The script begins as Luna transporter is T minus one hour, 30 minutes on the final countdown of the launch. Once launched, the four crew members aboard Luna will fly it to a landing at the Sea of Tranquility Base which has already been established on the moon. Enroute, the Mission Specialist goes outside the spacecraft to deploy satellites, which will orbit the moon. During the simulated flight, crew members wear spacesuits made from painting overalls they have decorated with space patches. There are members of the cast who portray five mission coordinators: launch and landing, mission control, public affairs officer, and navigator. It is suggested that slides of views of the Earth and the moon from space be projected as part of the background of the stage setting.

Stage Play: *Space Play—Tour of the Solar System* (from *Thematic Unit: Space* by Ruth Young)

This is an imaginary tour of the solar system aboard a futuristic spaceship, complete with crew members, tour guides, and passengers. The play begins with a launch from Earth. All the planets are visited along the way, and pictures are projected to add realism. The ship is bombarded by asteroids in the Asteroid Belt. The passengers on board hear the tour guide's narration regarding the planets they pass. While returning to Earth through the Asteroid Belt, one smashes the viewing porthole, but a shield automatically drops in place and all are safe.

Family Science •

Science Plays (cont.)

Stage Play: *Flying a 737* (from *Science Simulations: Intermediate* by Ruth Young)

This play was written by a commercial pilot. It simulates a flight of a 737 from Miami, Florida to Atlanta, Georgia. The script involves ground crew and flight crew, as well as passengers as they experience the flight. The excitement of a storm which threatens the plane's landing in Atlanta and a passenger who has symptoms of a heart attack add excitement and realism to the flight. Students become familiar with the flight terms used during the flight between flight and ground crew. Route maps should be used on an overhead projector to show the audience maps the pilots would use during this flight. Authentic representations of the instrument panels are included so they can be enlarged, and reproduced, and added to the cockpit on stage.

Shadow Puppet Show: *Life in an Ant Farm* (from *Science Simulations: Intermediate* by Ruth Young)

The cast of this puppet show are ants and a spider. Pictures of the cast and scenery are provided to be reproduced on transparencies and used on an overhead projector as in *The Story of a Honeybee* (see below). The story depicts the queen ant beginning the ant colony as she lays eggs and tends the larvae until there are enough of them to begin digging tunnels. They now take care of the queen, larvae, and pupae, including foraging to feed the members of the colony. While searching for food, some of the ants are stepped on by a human. Ants from another colony intrude on the nest and carry away some of the larvae. The play ends by closing the loop of the life cycle as a new queen ant uses her temporary wings to fly from the nest. She is followed by winged male ants that ultimately die after mating with her and are consumed by a spider.

Shadow Puppet Show: *The Story of a Honeybee* (from *Literature Unit: Magic School Bus Inside a Beehive* by Ruth Young)

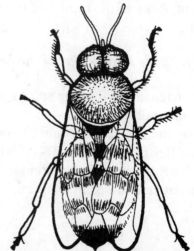

The cast for this puppet show are honeybees and a wasp. The book includes pictures of these cast members and scenery to reproduce on transparencies and use on an overhead, projecting the image onto a screen for the audience. The play tells the story of the honeybee from the points of view of the queen bee and workers as they carry out their lives in the hive and out in the fields. It portrays the life cycle of the bees, danger posed by an intruder in the hive (a wasp), and the search by a new queen and some workers for a new hive location.

Science Plays (cont.)

Reader's Theater: *Final Destination—Mars* (from *Science Simulations: Challenging* by Ruth Young)

Students read excerpts from a space traveler's journal, written as she goes to Mars from Earth in 2025 aboard the spacecraft Phobos. They begin their journey at the Space Station Freedom and are sent on a trajectory to Mars, a distance of millions of miles (kilometers). The traveler describes the effects of changes in the gravitational force as the spacecraft accelerates and then drifts in space. Views of Earth as the spacecraft travels farther from the planet are also described. Once at Mars, 28 days later, the traveler leaves Phobos on a shuttle capsule to the surface and the Jamestown II base. Views of Mars during the landing are described. It is suggested that slides and a computer-generated video of a flyover of Mars' surface (made by the Jet Propulsion Laboratory from the Viking mission data) be used to enhance this performance. Information on obtaining these materials is included in the book.

Reader's Theater: *Life of a Bat* (from *Stellaluna* by Janell Cannon)

This is a delightful story of a baby bat that becomes separated from her mother while flying away after an owl attack. Stellaluna drops into a bird nest and is fed bugs by the mother bird as she feeds her own young. The young birds think the new baby is very strange since she does not like bugs, and she hangs by her feet. The baby birds try this as well, only to be scolded by their mother. Stellaluna is told to behave or leave the nest. She tries hard to be just like the birds. Flying is no problem, but when she tries to land on her feet, she finds it nearly impossible. During one flight, she is reunited with her mother who takes Stellaluna to eat the fruit on which her species of bat lives. Stellaluna goes back to visit the birds in their nest. As she leaves that night, the three birds try to join her. They cannot see in the dark and begin to fall, so Stellaluna rescues them. They realize that they seem alike but are really different. They agree to remain friends, however.

Rewrite this story as a reader's theater script. Show the delightful illustrations as it is read aloud to an audience. The pictures may be projected by means of a video camera linked to a monitor. Bat and bird puppets may be made or purchased for use as a puppet show. These are available through a variety of science stores and catalogs.

Family Science Summary

Students, families, and school staff will all benefit from the experience they have at Family Science events. Families will benefit by having learned of the wide range of options open to them for experiencing science and learning more about the world in which they live. They will also be more likely to support staff efforts at teaching science through active participation. The overall benefit will be an improvement in the scientific literacy levels of all those involved.

Resources

Related Books and Periodicals

Bosak, Susan. *Science Is . . . A Sourcebook of Fascinating Facts, Projects, and Activities.* Scholastic, 1992. This huge, award-winning K–8 grade teacher's guide is packed with hands-on experiments covering a wide range of science topics. Available through NSTA. (See Related Materials and Organizations.)

Caney, Steven. *Steven Caney's Invention Book.* Workman Publishing, 1985. This book gives brief histories of famous inventions such as the Band-Aid, Life Savers, and Frisbee. It also suggests ideas for inventions children can make.

Cannon, Janell. *Stellaluna.* Harcourt Brace & Co., 1993. This delightful book tells about the struggle of a baby bat to fit into a family of birds when she is dropped into their nest by accident. The physical characteristics of birds and bats are discovered through the story, as well as a message about accepting differences in others.

Cobb, Vicki. *Chemically Alive! Experiments You Can Do at Home.* Lippincott, 1985. This outstanding book has simple-to-do chemistry experiments using household materials.

McCauley, David. *The Way Things Work.* Houghton Mifflin Co., 1988. This great book shows how, in a very humorous and reader-friendly manner, mechanical things work. It may inspire students to think of invention ideas. Also available on a CD-ROM.

National Science Education Standards. National Academy Press, 1996. This document contains the details of the K–12 National Science Education Standards adopted nationwide in 1996.

National Science Resource Center. *Resources for Teaching Elementary School Science.* National Academy Press, 2101 Constitution Ave., NW, Lockbox 285, Washington, DC, 20055, 1996. This outstanding resource book includes reviews and detailed information about curriculum materials, teacher's guides, science places to visit across the nation, and other sources of science materials.

National Science Teachers Association. *Science & Math Events: Connecting & Competing.* National Science Teachers Association, 1990. This book covers information on how to organize and conduct science events, including science fairs and invention fairs.

Sae, Andy, S.W. *Chemical Magic from the Grocery Store.* Kendall Hunt, ISBN 0-7872-2900-8. This book is filled with chemical experiments and teacher demonstrations. Each of these is clearly described, including background information, materials, cautions, procedure instructions, and discussion ideas for use with students after the activities. Most of these are appropriate for students in fifth grade and beyond.

Resources (cont.)

Related Books and Periodicals (cont.)

Science and Children, volume 34, number 2. National Science Teachers Association. October 1996. This issue of the NSTA elementary/middle school journal covers the topic "Families Involved in Real Science Together." Articles provide descriptions of a variety of family-oriented science programs which may be offered by schools. Order this publication through NSTA. (See Related Materials and Organizations.)

Sober, Ed. *Inventing Stuff.* Dale Seymour Publications, 1996. Designed for teachers and students to help them develop ideas for inventions.

Van Cleave, Janice. *Chemistry for Every Kid—101 Easy Experiments That Really Work.* John Wiley & Sons, 1989. Another great book in the series which has made this author so popular with teachers and students. Most of the easy-to-do experiments are appropriate for students in grades four and beyond.

Van Cleave, Janice. *Guide to the Best Science Fair Projects.* John Wiley & Sons, 1997. Covering a wide variety of science projects in the areas of astronomy, biology, earth science, engineering, and physical science, these projects are well-described and designed for a wide range of age groups.

Young, Ruth M. *Literature Unit: The Magic School Bus Inside a Beehive.* Teacher Created Materials, 1997. This unit includes a shadow puppet play, "The Story of the Honeybee."

Literature Unit: The Magic School Bus Inside the Earth. Teacher Created Materials, 1996. This book includes activities on identifying minerals, making simulated books, and walking through the rock cycle.

Literature Unit: The Magic School Bus on the Ocean Floor. Teacher Created Materials, 1996. This book includes an activity on dissecting fish.

Science is Fun: Primary. Teacher Created Materials, 1999. This teacher resource is filled with activities in the life, earth/space, and physical sciences for primary students. It includes activities with owl pellets, mixing liquid colors, simple chemistry, moon phases, mirrors, and pinwheels.

Science Simulations: Challenging. Teacher Created Materials, 1997. The activities in this book cover deciphering an alien message, travel to Mars in 2025, and a species evolution game.

Science Simulations: Intermediate. Teacher Created Materials, 1997. The activities in this book cover electric circuits using batteries, chemistry using cabbage extract indicator dyes, and flight using paper planes. Also included is a script for flying a 737.

Space: Intermediate. Teacher Created Materials, 1994. This book includes a play based on a simulated flight to the moon in the year 2021.

Thematic Unit: Space. Teacher Created Materials, 1998. This book includes the space play, "Tour of the Solar System." Activities on the history of astronomy and a study of the planets are included. Appropriate Web sites are suggested to support many of the activities.

Resources (cont.)

Related Materials and Organizations

AIMS (Activities Integrating Mathematics and Science) Educational Foundation. PO Box 8120, Fresno, CA 93747-8120 or www.AIMSedu.org/. Provides a variety of teacher guides on math and science activities for elementary grades.

CORE (NASA's Central Operation of Resources for Educators). Lorain County JVS, 15181 Route 28 South, Oberlin, OH 44074. Provides materials from NASA to use with students, including slides, videos, and photographs, as well as curriculum.

Delta Education, Inc. PO Box 3000, Nashua, NH 03061-9913 or www.delta-edu-com. Supplies a variety of science materials for K–8, including magnets, iron filings, and sun-print paper.

ETA Science Catalog. 620 Lakeview Parkway, Vernon Hills, IL 60061-9923, or e-mail at infoaetauniverse.com. Supplies a variety of science equipment, including mirrors and a 96-page book, *Mirrors*, with 50 easy experiments that use mirrors and flashlights.

Family Science, Northwest EQUALS, Portland State University. PO Box 1491, Portland, OR 97201-1491. This is a developer and national disseminator of Family Science program information. EQUALS supplies a book on implementing the program Family Science by P. Noone, 1995. Inquire about teacher-education workshops and teacher-scientist partnerships.

GEMS (Great Explorations in Math and Science). Lawrence Hall of Science, University of California, Berkeley, CA 94720. These are teacher guides on a wide range of topics integrating mathematics with life, earth, and physical sciences. The activities in these books foster a guided discovery approach to learning and are designed for grades K through 12. Topics include acid rain, bubbles, and a crime lab.

Genesis, Inc. PO Box 2242, Mount Vernon, WA 98723. This company sells owl pellets individually or in bulk.

Insights Visual Productions, Inc. PO Box 644, Encinitas, CA 92024. Request a free catalog and information regarding a videotape about an Inventors Showcase.

International Convention Services, 1645 N. Vine Street, Suite 611, Hollywood, CA 90028. Ask about Invention Convention information.

Invent America! PO Box 26025, Alexandria, VA 22313. This is a national education program for K–8 students, sponsored by U.S. Patent Model Foundation. "How to Get Started" kits are available for purchase by schools or families.

National Science Teachers Association (NSTA). 1840 Wilson Boulevard, Arlington, VA 22201 or www.nsta.org/scistore. The NSTA supplies science books and materials for K–12 teachers. Request a publications catalog.

Scientific (Edmund Scientific Co.). 101 East Glouchester Pike, Barrington, NJ 08007-1380 or www.edsci.com. This company supplies a wide range of science materials and toys, including magnets, iron filings, and sun-print paper.